高蝉多远韵
茂树有余音

赵志模教授八十华诞纪念
学术文集选

王进军 ◎ 主 编

西南大学出版社
国家一级出版社 全国百佳图书出版单位

图书在版编目(CIP)数据

高蝉多远韵,茂树有余音:赵志模教授八十华诞纪念学术文集选/王进军主编. -- 重庆:西南大学出版社,2024.12. -- ISBN 978-7-5697-2824-8

Ⅰ.Q968.1-53

中国国家版本馆CIP数据核字第2024PN1677号

高蝉多远韵,茂树有余音
——赵志模教授八十华诞纪念学术文集选

GAOCHAN DUO YUANYUN, MAOSHU YOU YUYIN
——ZHAO ZHIMO JIAOSHOU BASHI HUADAN JINIAN XUESHU WENJI XUAN

主　　编	王进军
责任编辑	杜珍辉　徐庆兰
责任校对	邓　慧
装帧设计	汤　立
出版发行	西南大学出版社(原西南师范大学出版社)
	地址:重庆市北碚区天生路2号
	邮编:400715
	市场营销部电话:023-68868624
印　　刷	重庆升光电力印务有限公司
成品尺寸	170 mm×240 mm
印　　张	18.5
插　　页	28
字　　数	380千字
版　　次	2024年12月　第1版
印　　次	2024年12月　第1次印刷
书　　号	ISBN 978-7-5697-2824-8
定　　价	88.00元

编委会

主　编

王进军

执行主编

何　林

副主编

丁　伟　邓永学　刘　怀　张永强

编　委（按姓氏笔画排序）

丁　伟　王文琪　王进军　尹　勇　邓永学　叶　辉　叶鹏盛
吕龙石　吕慧平　刘　怀　刘　洪　闫香慧　李剑泉　杨　洪
邱明生　何　林　何恒果　沈　丽　张永强　张肖薇　张建萍
张智英　陈　宏　陈　艳　陈文龙（贵州）　陈文龙（美国）
武可明　范青海　罗华元　金道超　周亦红　孟庆繁　赵云芬
宫庆涛　郭凤英　唐　松　涂建华　曹小芳　程绪生　薛传华

编写说明

鉴于历史变迁,发表论文时的学术标准与当下相比存在一定差异。在经过与原作者的沟通协商后,我们对论文进行了精细的审校,除对文中少数细节进行了必要的修正外,尽量保持了论文的原有风貌。以下是对相关处理措施的详细说明:

1.鉴于各篇论文最初发表于不同期刊,我们尊重原文的摘要、关键词和标题,仅在格式上进行了微调,对于缺失的部分并未进行补充。

2.在论文发表的年代,部分物种的名称及拉丁学名是符合当时的分类标准的。然而,随着分类学的不断发展,这些名称可能已经经历了多次修正。在本书中,我们尽量保留了原文的表述,仅对一些明显的错误进行了更正。

3.在字词的使用、量的表述以及单位的标注上,我们尊重了历史的原貌,如"桔"(橘)、"mM"等表述均未做统一修改。对于数据存在疑问的地方,我们优先采用了作者原文的表述,仅对个别的错误进行了修正。

4.由于书稿篇幅的限制,论文的参考文献未能收录于本书中。如读者有进一步查阅的需要,建议通过互联网下载原文以获取完整的参考文献信息。

序 一

师 恩 难 忘

在即将迎来恩师赵志模先生八十华诞之际，作为第一个进入赵志模先生门下的博士后研究人员，回想起追随先生学习和研究的岁月，先生广博的学识、严谨的教学和科研态度、对学生无微不至的关怀和指导仍历历在目，至今难忘。

我对先生最初的了解和景仰缘起于先生编著的《生态学引论——害虫综合防治的理论及应用》和《群落生态学原理与方法》两部当时国内系统介绍生态学理论与方法的权威著作。我博士研究生的选题是关于东北林区生物多样性方面的研究，当时国内群落生态学和生物多样性研究刚刚起步，相关书籍和参考资料较少，先生的著作对我博士研究工作设计和数据解析帮助巨大。其后，先生作为我博士学位论文的评阅人之一，在评阅意见中对我的论文给予了充分肯定，同时对论文中存在的问题提出了详细的修改意见，这进一步加深了我对先生渊博学识和严谨科研态度的景仰和敬意，这份评阅意见我至今仍珍藏着。

1996年6月，我顺利地通过了东北林业大学博士学位论文答辩。和当时不少毕业博士研究生一样，仍想通过博士后研究进一步开阔自身的学术视野，进而提升科研能力。出于对群落生态学、生物多样性研究的兴趣和对先生学识的由衷景仰，我鼓足勇气给先生写信表达想拜在先生门下继续深造的愿望，先生很快回信表示欢迎。当年7月上旬，我专程赴重庆拜见先生。第一次面见先生这样的生态学大家，还是心存忐忑，但见面交谈后，先生的亲切谦和给我留下深刻的印象。我详细汇报了我的学习和科研工作情况，表达了追随先生从事博士后研究的强烈愿望和科研工作设想，先生支持我申请进入西南农业大学农学博士后科研流动站，并同意作为合作导师指导我的博士后研究工作。至此，我有幸正式进入师门。

当年9月份，我与妻子来到西南农业大学，追随先生开始了为期两年的博士后研究工作。在学习和研究工作期间，有更多的时间和先生相处，有更多的机会聆听师训和向先生请教，先生严谨的治学态度、孜孜以求的敬业精神时刻激励我在学术之路上前行。就我博士后工作期间的科研选题，先生与我多次讨论，考虑到我已有的研究基础和重庆缙云山生物多样性研究情况，建议我开展缙云山节肢动物多样性方面的研究，并对研究方案的制定给予了细心的指导和帮助。缙云山地处亚热带，植被类型以常绿阔叶林和杉木林为主，与温带落叶阔叶林和针阔混交林植被组成差异巨大。不熟悉缙云山森林植被，不认识亚热带森林植物物种，给我的研究带来很大挑战。先生了解情况后，与西南师范大学联系，帮助我找到了一名生态学硕士研究生协助森林植物调查，解决了研究中遇到的最大难题。同时先生还安排了欧庆宇等同学协助开展外业调查和内业整理等工作，请求朱文炳先生帮助鉴定蜘蛛标本，这些为我的研究工作顺利开展提供了充分的支持。

印象深刻的是先生领导下的昆虫生态学实验室和谐、进取的团队精神，内容广泛的科学研究以及活跃的国内外学术交流活动，耳濡目染，沉浸其中，我的学术视野得以开阔、我的科研协作精神得以养成。

另外，先生和冷师母对我和妻子生活上的关怀也令我感动不已。我生长和求学在东北地区，重庆夏季的炎热、冬季的寒冷以及饮食的辛辣也给我们的生活带来了种种"挑战"。先生帮助我们安排了舒适的居住条件，解决了气候不适应难题。先生多次请我们到家中做客，冷师母亲手为我们烹制最具特色的好吃的重庆菜肴，使我们慢慢地接受和喜欢上了重庆菜。

先生的关怀使我愉快地度过了两年紧张和收获丰厚的博士后研究工作时光，顺利地完成博士后研究工作。

这些虽然仅是我在先生身边学习和工作的几个片段，却是我一生难忘和备感珍惜的记忆。谨以此恭祝先生八十华诞生日快乐，仁者长寿！

孟庆繁

序 二

师 缘 小 记

师弟刘怀来电来信告知要为导师赵志模教授八十寿诞举办俭朴纪念活动,才幡然回首,真是时光如梭啊,不觉间竟然已脱离昔日西南农业大学博士研究生身份29年了!由于工作关系,时常会回到如今的西南大学,也就常有机会亲近美丽的校园,常有机会与老师和在校师兄师弟师妹小聚,或许如此,才令自己忽略了似水流年。

记得是1992年春夏时节,经郭依泉师兄引荐报考李隆术先生的博士研究生,为考好,在时西农干训部招待所准备功课,其间,偶有机会在依泉的导引下拜识了赵老师,因之前读过赵老师的著作,初次见面即在心里建立了赵老师文如其人的亲切感。老师论著之科学严谨,自不待言,然文体文风文字之细腻、平实、耐读、易解,有如赵老师与学生的相处之道——用心微至,严而不苟,门下众多弟子或许比我体会更加深刻。

十分幸运地顺利通过了博士研究生初试,1992年9月入学西农,由于之前得到"中英友好奖学金"资助赴英做访问学者,已计划于1993年1月启程。到西农报到后,向李隆术先生面陈论文选题——鉴于国内水螨分类研究几为空白,加之前期自己做了一些铺垫工作,拟选题水螨分类学研究,李先生欣然认可。交流中李先生给予了许多语重心长、推心置腹的教诲,因时已久远,嘱语难详,但要点至今在心,其中之一大意是:不管选题做什么,研究工作是什么领域,做学问要尊重科学思维的基本规律,赵志模老师就是深谙学问之道的典范,以后多接触。李先生轻描淡写"多接触",我却已领会并认定,先生意指,赵老师也是我的导师。当时便想,与赵老师初次见面时内心油然而生的亲切感,原来是有因缘的。

从英国回归后,1994年9月复入西农校园至1995年7月,补课程,赶论文,皆得赵老师传教。紧张之余,和师弟师妹们常有逗趣小聚,赵老师有邀必至,教研室老师组织活动,赵老师等也常把弟子们捎上,耳濡目染,赵老师为师秉性潜移默化植根于心。归在西农学习生活所得,概为一句:时听赵老师授业可深悟教习之法,时与赵老师小聚可深悟为师之道。对此,师姐师兄师妹师弟在文集里定然有许多学习生活的故事旁证。因此,这篇短文只简记与赵老师的结缘。

赵老师是学业导师,更是人生导师。幸甚!

金道超

序 三

师 生 情

时间如白驹过隙,距离考上西农的那天,已过去31年了,至今想起,仍然心潮澎湃!回忆起那些在西农求学的峥嵘岁月,更是热血沸腾!

那是1993年秋天,我怀着求知的欲望、人生的梦想、激动的心情来到了祖国的大西南重庆北碚——西南农业大学,踏上了我人生转折的征程。

母校的山山水水,一草一木至今仍记忆犹新,矗立在校园中植保系的那座山峰,我和师弟师妹们喜欢亲切地称它为"小山坡",在那里留下了老师们教诲的声音、留下了我们登山求学的脚步声、留下了科学研讨的争论声、留下了生活快乐的谈笑声。研究生的三年经历已深深地刻在我的脑海里,难以忘怀导师赵志模教授在学业上给我们的谆谆教诲、在科学实验技能上手把手悉心的指导;难以忘怀导师在生活上无微不至的关怀,仍记得重庆的七月,酷暑炎炎,在导师家避暑,犹如自己的家;寒冬腊月在导师家避寒、过年,犹如自己的家;平日里我们的学生宿舍常常留下导师的足迹、问寒问暖,也如自己的家……学习的三年,无论在田间调查经受日晒雨淋还是在室内日夜检索进行标本鉴定都没有一丝辛苦的感觉、没有困难的感觉、没有远离他乡的感觉。这三年我很快乐,很幸福,很向往,很留恋……这是因为在我们的后面有我们敬爱的导师、亲爱的父亲——赵志模先生!

"滴水之恩,将涌泉相报"这是导师对我语重心长的教导,也是导师对我道德水准的要求,为我人生道路指明了方向。毕业后的这些年无论在科研上、教学上、管理上还是做人上受益于导师的传授,秉承导师的教导为社会为人民默默贡献。在导师八十寿诞之际,心情无比激动,导师一辈子为我们呕心沥血、无私奉献,千言万语难以表达感激之情,在此,我对我的导师表示衷心的感谢!感谢恩师给我们前进的动力;感谢恩师给我们飞翔的翅膀;感谢恩师给我们指明人生的方向;感谢恩师给我们放眼世界的慧眼。我的一言一行没有愧对导师的教诲!请导师放心,我会一如既往为祖国贡献余力!

陈宏

前　言

《高蝉多远韵,茂树有余音——赵志模教授八十华诞纪念学术文集选》一书原本是赵志模教授的弟子们在他八十寿辰上敬献的生日礼物并准备出版的,但由于疫情等因素耽误了,五年后的今天终于正式出版了。

赵志模教授于1939年4月18日出生于四川省涪陵地区南川县(今重庆市南川区)。20世纪30年代是一个动乱的年代,是中国人民抵御外国列强入侵、生活处于水深火热的年代,赵志模教授的父亲曾是抗日战争中征战缅甸等地的远征军军人,待1945年退役返家时,赵志模教授已是一名小学二年级学生。赵志模教授一生的"三农"情缘与他的两个重要人生转折点有关:1950年3月,他考入南川中学,自愿转入农班就读,自此他一生的学习、工作没有脱离一个"农"字,也从此开始接触农作物病虫;1978年,他以考分第一的成绩考取了西南农学院的硕士研究生,师从著名昆虫学家李隆术等教授,自此走上了植物保护科学研究和教书育人的不悔之路。

赵志模教授1960年参加工作,2009年退休,实际科研工作和协助指导研究生至2014年,其中自1981年研究生毕业留校任教30余载。在50余年的教学、科研和推广工作中,他数十年如一日,勤勉耕耘,求真务实,在昆虫生态学与害虫综合治理、农产品储运保护、螨类毒理学研究领域成就斐然,为我国植物保护事业培养了大批优秀人才。

赵志模教授学术造诣精深,是国内外著名的昆虫学、生态学专家,他出版的《生态学引论——害虫综合防治的理论及应用》和《群落生态学原理与方法》两本姊妹篇专著曾是从事昆虫生态学与害虫综合治理研究的科技人员和在校研究生们的案头必读之物,这也奠定了他在害虫生态学理论与应用研究领域的重要地位。他曾任西南农业大学植物保护系主任、重庆市植物保护学会理事长、重庆市生态学会副理事长、重庆市农药协会副理事长、重庆市农学会常务理事、中国植物保护学会

常务理事、中国植物保护学会科学普及工作委员会副主任、中国昆虫学会蜱螨专业委员会委员、国务院学位委员会学科评议组成员、全国高等农业院校教学指导委员会农业昆虫小组组长。担任过《植物医生》主编,Systematic & Applied Acarology、《昆虫知识》和《西南农业大学学报》编委等职。赵志模教授科研成就卓著,先后获农牧渔业技术改进一等奖1项、科技进步三等奖4项,四川省科技进步一等奖1项、二等奖3项、三等奖1项,重庆市科技进步二等奖1项、三等奖1项、自然科学三等奖1项;发表中英文学术论文320余篇,出版专著、编著或主编教材9本;获授权专利5项。执教30余年,先后为本科生、研究生讲授害虫生物防治、农业昆虫学、昆虫生态与预测预报、昆虫生态学、害虫种群系统控制、昆虫学研究进展、理论生态学、群落生态学、系统分析在昆虫学中的应用、农药毒理学等10余门课程。他的课堂因其渊博的知识、深入浅出的讲解、幽默风趣的语言、平易近人的风格,深受学生喜爱。

赵志模教授从教30余载,以身作则,呕心沥血,培养了数十名硕士和博士,如今他们在各自工作岗位都成长为领导、业务骨干、学术带头人或企业家,他们所取得的每一点成绩都离不开赵志模教授曾经的悉心栽培。农牧渔业部部属重点高等农业院校优秀教师、重庆市教委为人师表先进个人、重庆市优秀博士生指导教师和西南农业大学优秀教师等荣誉称号正是对他"春蚕到死丝方尽、蜡炬成灰泪始干"的大先生精神和"春风化雨、润物无声"的诲人情怀的褒奖。赵志模教授作为教师,他一直坚持教书育人的理念,严谨治学,谦虚谨慎,严以律己,宽以待人;他把学生当成自己的亲人和同事,没有任何做老师的架子,又不失做教师的威严;他以身作则,谆谆诱导,教他们做人,扶他们成长,授他们知识,励他们成才;他尊敬先辈,奖掖后学,待人以诚;他有极强的团队精神,心怀开阔,从不计较个人得失。赵志模教授高风亮节的高贵品质、为人处世的生活态度以及他在教学、科研上取得的成绩,使他在全国植保学界受到尊重。

这本文集除选编了赵志模教授不同时期的代表性学术论文外,还涵盖其求学经历、职业生涯、学术成就、人才培养和工作掠影等内容,从中能强烈感受到他的学者风范、先生品格和君子气度。实际上,这本文集也可以看作是一本思政教材,在全社会倡导弘扬大先生精神、加强师德师风建设的当下也极具现实意义。

我和赵志模教授曾经是涪陵农校的校友,看到他如今已是学术有成,桃李满天下,很是为他高兴。祝赵志模教授健康长寿,安享晚年!

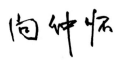

中国工程院院士、原西南农业大学校长

目录

赵志模教授简介 ··· 1
    第一部分　求学与工作经历 ··· 1
    第二部分　学术思想与成就 ··· 4
    学习与工作简历 ·· 14

论文选集 ··· 15
    主题一　昆虫生态学与害虫综合治理研究 ································· 15
        昆虫寄生作用数学模式的探讨 ·· 16
        桔园昆虫群落演替初步研究 ··· 31
        重庆市郊不同种植制度菜地昆虫群落结构研究 ····················· 43
        柑桔矢尖蚧一代幼蚧发生期数理统计预测 ···························· 53
        舞草种子的蚂蚁传播 ·· 61
        多物种共存系统中拟水狼蛛对三种稻虫的捕食作用 ··············· 72
        瓜类蔬菜昆虫群落(包括蛛形纲和软体动物)的研究Ⅲ.群落的数量动态 ··· 80
        不同微生境中泽兰实蝇寄生对紫茎泽兰有性繁殖的影响 ········ 91
        温度、土壤含水量和埋蛹深度对柑橘大实蝇羽化的影响 ········ 99
        柑橘大实蝇羽化出土及橘园成虫诱集动态研究 ···················· 108
        酸雨对植物—害虫—天敌系统的作用 ································· 117
        温度对竹盲走螨实验种群生长发育与繁殖的影响 ················· 124

目录 CONTENTS

主题二　农产品储运保护研究……………………………………………………135

　　双低储粮虫螨群落组成研究……………………………………………………136

　　腐食酪螨对低氧高二氧化碳气调的抗性………………………………………139

　　磷化氢熏蒸处理对嗜卷书虱不同虫态的致死作用……………………………143

　　CO_2和溴氰菊酯在不同温度下对嗜卷书虱毒性的相互影响………………152

　　Accumulation and Utilization of Triacylglycerol and Polysaccharides in *Liposcelis bostrychophila*(Psocoptera, Liposcelididae) Selected for Resistance to Carbon Dioxide……………………………………………………………………………159

　　Development and Reproduction of Three *Euseius* (Acari:Phytoseiidae) Species in the Presence and Absence of Supplementary Foods………………………………169

　　Toxicological and Biochemical Characterizations of GSTs in *Liposcelis bostrychophila* Badonnel(Psocop., Liposcelididae)…………………………………181

　　Infection by *Wolbachia* Bacteria and Its Influence on the Reproduction of the Stored-Product Psocid, *Liposcelis tricolor*……………………………………193

主题三　螨类毒理学研究……………………………………………………………201

　　桔全爪螨(*Panonychus citri* McGregor)实验种群生命表的组建与分析………202

　　朱砂叶螨对不同农药抗药性发展趋势的研究…………………………………214

　　温度对朱砂叶螨二种抗药性品系发育和繁殖的影响…………………………220

　　朱砂叶螨阿维菌素抗性品系选育及适合度研究………………………………227

　　黄花蒿提取物对柑橘全爪螨的生物活性………………………………………234

　　甲氰菊酯和阿维菌素对柑橘全爪螨的亚致死效应……………………………242

Effects and Mechanisms of Simulated Acid Rain on Plant-Mite Interactions in Agricultural Systems. I. The Direct Effects of Simulated Acid Rain on Carmine Spider Mite, *Tetranychus cinnabarinus* ·················252

Calculation of Developmental Duration of Mites Reared in Groups Compared to Those Reared in Isolation ·················261

附录一　专家荣誉称号、获奖证书和授权专利目录·················269

附录二　专著、教材、论文目录·················272

附录三　荣誉和获奖证书(部分)·················297

附录四　工作生活掠影·················305

附录五　指导的学生·················320

后记·················330

赵志模教授致编写组的亲笔信·················332

赵志模教授简介

第一部分　求学与工作经历

知名昆虫生态学研究者、农作物害虫综合治理践行者、西南大学二级教授、博士生导师赵志模先生，1939年4月18日出生于四川省涪陵地区南川县(今重庆市南川区)。其父曾是抗日战争中征战缅甸等地的远征军军人，1945年农历腊月退役返家，是年赵志模先生已是南川县隆化镇小学二年级学生。1950年3月，他考入南川中学。当时的南川中学由新中国成立前的南川国立县中、南川师范职业学校和南川农业职业学校合并而成。由于家庭经济拮据，他自愿转入国家包伙食、包分配的农班就读。这是赵先生求学和职业生涯中的第一个重要转折，自此他一生的学习、工作没有脱离一个"农"字，也从此开始接触农作物病虫。1951年全国中等专业学校整合调整，南川中学农班并入涪陵农校。此时赵先生年方12，他随同全班13位同学，在农科启蒙教师李厚泽先生(金陵大学毕业)带领下，穿山越岭，步行3天抵达涪陵。1953年3月，赵先生从涪陵农校毕业，随同全班同学赴成都四川省农业厅等待工作分配。由于他年龄尚小，还是少先队员，厅里让他回原校继续读书。当年全国大、中专改为秋季招生，他只好附读涪陵农校农产制造专业高二年级。同年9月四川省农业厅重新分配，赵先生独自一人赶乘载货木船，从涪陵顺流而下，历经两天抵达万县农校，就读植物保护专业。1956年9月从万县农校毕业，因成绩优异，被学校保送至西南农学院植物保护专业继续深造。1960年9月大学毕业后由国家统一分配至凉山州农科所工作。在此期间曾任植保研究室主任，从事小麦锈病、地老虎、黏虫、油菜蚜虫等病虫的研究。1962年底因照顾夫妻关系，调至越西县农业局，不久后成为该县第一任植保站站长，在全县农作物病虫预测预报和防治中加深了对植保工作重要性的认识，积累了基层植保工作经验。在越西县工作的16年间，他先后就职于农业局办公室、政府农办、县委农工部。

1978年，国家恢复研究生招生，赵先生以考分第一的成绩成为西南农学院李隆术、王辅、朱文炳教授共同指导的硕士研究生。这是赵先生职业生涯中的又一重大转折，自此他走上了植物保护科学研究和教书育人的不悔之路。在读研和其后的工作期间，他多次得到了中国科学院生态学研究中心马世骏、蓝仲雄先生，中国科学院动物研究所丁岩钦、李典谟先生，中国农业科学院植物保护研究所郭予元先生，复旦大学忻介六先生，华南农业大学庞雄飞先生，福建农学院（今福建农林大学）赵修复先生，北京师范大学徐汝梅先生，以及日本北海道大学森樊须先生等的帮助、鼓励和支持。赵先生还两次参加全国农业害虫天敌资源调查培训班，聆听了华南农业大学庞雄飞、福建农学院（今福建农林大学）赵修复、浙江农林大学何俊华等著名专家的讲课，这奠定了赵先生毕生从事昆虫生态学研究和践行害虫综合治理的思想基础。在李隆术教授等的悉心指导下，他以稻田寄生性天敌资源调查和稻田寄生蜂群落研究为题，顺利完成了学位论文撰写和学位论文答辩，成为"文革"后首批授予的农学硕士。复旦大学忻介六先生在书面评审其学位论文中，给出了"该硕士学位论文已达到博士学位论文水平"的评价。

1981年10月，赵志模先生硕士研究生毕业后留校任教，1986年晋升副教授，1990年晋升教授，1992年由国务院学位委员会批准为博士生导师，2007年评为二级教授，2009年退休，但完成科研项目和指导研究生的工作持续到2014年。1987—1988年曾在美国加州大学（UC）昆虫系进修和合作研究，师从世界著名农螨学家J. A. McMurtry教授。

在教学上，赵志模先生先后为博士生讲授"理论生态学""群落生态学""系统分析在昆虫学中的应用"，为硕士生讲授"昆虫生态学""害虫种群系统控制""昆虫学研究进展"，为本科生讲授"害虫生物防治""农业昆虫学""昆虫生态与预测预报"等多门课程。分别主编、副主编《农产品储运保护学》和《昆虫生态学与害虫预测预报》两本全国高等农业院校通用教材，参编《农业昆虫学》、《植物检疫》和《农业昆虫学实验指导》3本校内使用教材。"柑桔害虫多媒体课件"和"光温湿控制培养室的组装与使用"分获重庆市优秀课件二等奖和学校优秀教学成果二等奖。独立或合作培养、指导了22名硕士研究生、16名博士研究生，多次被评为农业部、重庆市和学校优秀教师。

在科研上,赵志模先生曾主持或主研各级各类科研课题20余项,其中国家自然科学基金项目"桔园生态系统及其控制研究""菜地昆虫群落及其控制机理研究""抗药性朱砂叶螨生态适应度及抗药性风险评估研究""阿维菌素抗性朱砂叶螨酯酶基因的克隆及表达研究""书虱对气调抗性的生化机理研究""朱砂叶螨适应酸雨胁迫的机理及生态适合度研究"等共7项;国家公益性行业(农业)科研专项课题"西南地区柑橘大实蝇监测与防控技术研究与示范"1项,国家"七五"、农业部"八五""九五"重点攻关课题"以生防为主的果树病虫综合治理研究""茶果病虫预报及优化防治措施研究"等3项;教育部博士基金"酸雨胁迫对作物—害螨—天敌的影响及作用机制研究"等2项;四川省、重庆市重点课题"四川省农业害虫和天敌资源调查研究""桔园昆虫群落及系统控制研究""玉米昆虫群落及害虫控制研究"等10余项;联合国粮食及农业组织(FAO)(以下简称联合国粮农组织)项目"改进四川农户粮食产后技术""水稻IPM技术的应用"等2项。此外还被3个"973"国家重点基础研究项目聘为学术顾问。出版《生态学引论——害虫综合防治的理论及应用》《昆虫群落生态学》《四川农业害虫天敌图册》《四川农业害虫及其天敌名录》等专著4本;在国内外相关学术刊物上发表论文《稻田寄生蜂群落种-多度关系、多样性及群落排序的探讨》《昆虫寄生作用数学模型探讨》《重庆市北碚区稻田寄生蜂类群初步考察》《桔全爪螨 Panonychus citri McGregor 实验种群生命表的组建与分析》《重庆市郊不同种植制度菜地昆虫群落结构研究》《柑桔矢尖蚧一代幼蚧发生期数理统计预测》、Calculation of developmental duration of mites reared in groups compared to those reared in isolation、Development and reproduction of three Euseius (Acari:Phytoseiidae) species in the presence and absence of supplementary foods 等320余篇,其中国外刊物和国内用英文发表的论文47篇,SCI源刊物18篇。"四川省农业害虫和天敌资源调查研究""柑桔叶螨种群生态系统研究""桔园生态系昆虫群落及其控制""柑桔园昆虫群落及害虫综合治理研究""柑桔害螨综合管理决策支持系统研究""柑桔病虫综合防治新技术研究""菜地昆虫群落及害虫综合防治研究""四川省主要农作物病虫害抗药性监测与研究""新传入检疫对象——美洲斑潜蝇的研究与防除"等研究成果分获农业部、四川省和重庆市科技进步一等奖3项,二等奖4项,三等奖6项,重庆市自然科学三等奖1项。

赵志模先生曾任西南农业大学植保系主任,农业部昆虫学及害虫控制工程重点实验室主任,国务院学位委员会学科评议组成员,全国高等农业院校教学指导委员会农业昆虫小组组长,重庆市植物保护学科学术技术带头人,四川省和重庆市科技顾问团成员,重庆市第十届政协委员,享受国务院政府特殊津贴。

第二部分 学术思想与成就

赵志模先生在长期的学术生涯中,一直坚持理论联系实际和学以致用的原则,强调以生态学的理论和方法、系统论和整体观的思想来研究和指导害虫治理。他的学术思想和学术成就集中体现在昆虫生态学与害虫综合治理、农产品储运保护和螨类毒理学研究三个方面。

1.昆虫生态学与害虫综合治理研究

20世纪60年代以来,食物、人口、能源、自然资源的开发利用、环境保护等世界性五大社会问题日益突出,引起人们严重关注。这五大社会问题无不与生态学密切相关。面对实践的挑战,国内外一大批杰出的学者纷纷致力于生态学的研究。与此同时,现代科学的三大理论支柱系统论、控制论、信息论的日臻完善和向各门学科的渗透,为生态学提供了崭新的理论和研究手段。正是在此背景下,赵先生从20世纪70年代末开始接触生态学,并于80年代初由上海复旦大学忻介六先生推荐,获批了其首个中国科学院科学基金(国家自然科学基金的前身)资助项目"系统生态应用理论研究"。在导师著名昆虫学家李隆术教授的指导下,赵先生广泛涉猎生态学论著,博采众家之长,同时结合自己的研究成果及独到见解,先后以学院的名义举办了两期共40天的"西南地区昆虫生态学及害虫测报培训班"。在培训班讲稿的基础上,赵志模先生和周新远先生合作,于1984年出版了《生态学引论——害虫综合防治的理论及应用》专著。这本关于种群生态学的著作堪称经典,创造性拓展了我国昆虫生态学及害虫综合治理的理论与实践,充分体现了赵先生在昆虫生态学上的学术造诣和创新精神,这本专著也成了当时从事昆虫生态学与害虫综合治理研究的科技人员和在校研究生们的案头必读之物。

赵先生及其学术团队在1985年第三期《西南农学院学报（系统生态应用理论研究专辑）》上发表的研究论文，涉及柑橘叶螨种群的抽样策略及密度估计、种群动态的系统分析及其模拟、种群的空间格局及扩散聚集趋势、种群的密度效应及调节机制、种群生命表的组建与分析、害虫的危害损失与多因子经济阈值、种群系统模型组建及最优化管理等种群生态学研究的方方面面。这些论文也较为全面地反映了赵先生及其团队在种群生态学领域的学术思想和研究成果。

《生态学引论——害虫综合防治的理论及应用》一书开篇指出，现代生物学向两个方向迅猛发展，一个是微观，一个是宏观。分子生物学、细胞生物学向微观方向发展，在细胞水平、分子水平探索着生命现象的秘密；生态学则向宏观方向发展，在种群、群落、生态系统，乃至生物圈的水平上揭示生命系统的奥秘。成书至今的40年间，这一论述已经得到证实。该书首次提出了"种群系统"的概念并对其内涵作了深入剖析：种群由同种有机体的个体组成，但绝不等于个体的简单相加，从个体到种群是一个质的飞跃。种群具有可与个体相类比的一些特征，例如个体是生或是死、是雌或是雄、寿命有多长、是什么虫态或虫期、是否处于休眠或滞育状态等，但是作为由个体组成的种群，个体的这些生物学特性则是由统计特征表现出来的，即种群的出生率、死亡率、性比、平均寿命、年龄结构、休眠或滞育百分率等。同时，种群作为一个群体结构单位，还具有一些不为个体所具有的特征，如种群密度、数量动态、空间格局、聚集扩散以及种群的遗传特征；在自然界中种群是物种存在、物种进化、能量流动及物质循环的基本单位，是生物群落及生态系统最基本的组成成分，也是生物资源利用和有害生物治理的具体对象；种群内个体之间、种群与其环境之间的相互关系构成了种群系统，这个系统同其他系统一样，具有信息传递、行为适应、反馈控制的功能。针对以往种群动态研究中存在数量动态和空间分布两个分支泾渭分明的现象，赵先生创造性地提出，在深入探讨种群变动的机制和揭示害虫猖獗发生的原因时，应该从统一观出发，将种群的数量、空间和时间作为一个联合的结构系统。种群数量是系统的本质，没有数量就失去了研究的意义；种群需要也必须占据一定的有效空间,以保证种群的正常生活和繁殖;种群的生活繁殖也需要且必须有一定的时间。时间的连续性保证种群完成正常的生长发育和生活周期,时间的阶段性,使种群在时间序列上呈现出不同的数量特点和空间格局。赵先生的《生态学引论——害虫综合防治的理论及应用》一书中,在多次参加的国内、国际会议上都坚持这样一个观点,即有害生物的综合防治或综合治理必须以生态学、经济学和环境保护思想作为指导,以优化生态系统作为管理目标,充分利用自然控制因素,把有害生物的数量控制在经济允许水平之下。在生态学研究方法上,赵先生及其学术团队创造性地应用模糊数学、灰色系统等描述害虫种群发生规律;应用系统分析和箱车法模拟种群动态并建立种群最优管理的计算机决策支持系统;应用蒙特卡洛模拟分析种群的分布格局,利用平均密度和平均拥挤度的时序变化阐明种群在田间的聚集与扩散趋势,所有这些都表现了赵先生在昆虫生态学领域勇于创新的精神和透过现象看本质,抓住事物的本质特征来研究问题的能力。

赵先生认为种群生态学是以特定害虫为中心的综合治理的重要理论基础,但他在研究和实践中发现,针对单一害虫制定的任何一条综合防治措施,都会对农田生态系统中的各种生物成分,包括农作物、田间杂草、目标害虫、次要害虫、中性昆虫、天敌昆虫以及致病微生物等带来不可估量的影响,以单一害虫为目标的综合治理,并不能解决害虫再"增"猖獗、次要害虫上升等问题。因此建立以作物为中心、以农田生态系统为目标、针对多种害虫的综合防治体系不仅是生产实践的需要,也是害虫综合防治自身发展的必然结果。基于"群落的发展而导致生物的发展"这一基本观点,昆虫群落生态学应该成为以作物为中心的害虫综合治理的重要理论基础。

群落是较个体、种群更高一级的组织层次,是连接种群和生态系统之间的桥梁。群落生态学是现代生态学理论中重要的组成部分。它是集中研究生态系统中有生命的部分——生物群落的科学,其目的在于了解群落的起源、发展、各种静态和动态特征以及群落内及群落间的相互关系,从而深化对自然界特别是对生态系统的认识,为人类充分而合理地利用自然资源,提高生态系统生产力,推动生物群落向特定方向发展,保持生态系统的相对稳定与平衡提供理论依据。

赵志模先生的研究生毕业论文就是有关昆虫群落生态学的研究。论文首先对重庆市北碚区稻田的寄生蜂名录进行整理，分析了不同稻型田寄生蜂类群的组成和数量消长规律，探讨了稻田主要害虫、主要寄生蜂和重寄生蜂所形成的食物链和食物网；研究了稻田寄生蜂群落的种-多度关系、多样性以及群落排序，提出了不同耕作制度稻田寄生蜂群落沿环境梯度变化的速率和群落异质性水平的指标。最后，以纵卷叶螟绒茧蜂和稻纵卷叶螟为例，探讨了昆虫寄生作用的种群模式。研究生阶段的研究不仅奠定了赵先生从事群落生态学研究的基础，同时也增强了他对该领域研究的兴趣。从20世纪70年代末开始，在国家自然科学基金、国家"七五"攻关课题、四川省科委重点课题的资助下，赵先生在群落生态学方面作了大量研究工作。对昆虫群落的组成和食物网结构，群落的演替、时间和空间格局，生态位，竞争群，多样性指标等方面的研究均有不少新的创见，也探索出不少新的研究方法。1988年第二期《西南农业大学学报（桔园昆虫群落及害虫综合治理研究专集）》的出版和其他先后发表的多篇有关群落研究的学术论文，在国内外引起较大反响。赵先生被中国农科院茶叶研究所、棉花研究所邀请，就昆虫群落生态学的原理和方法前去讲学一周。鉴于当时国内还没有一本系统的、理论和方法兼顾的群落生态学专著，赵先生参阅了国内外大量文献资料，结合自己一系列科研成果，在对外讲学时编撰的讲稿基础上，与郭依泉先生合作，于1990年出版了他的第二本专著也是全国第一本《群落生态学原理与方法》生态学专著。该书反映了群落生态学发展的过去、现在和未来，系统阐述了群落生态学的基本原理、群落食物网随机理论、群落中的种间关系、群落丰富度及种面积、种-多度关系、群落的多样性及稳定性、群落排序及分类等。该书再次体现出赵志模先生勇于创新的精神和理论联系实际的学术思想。

群落食物网是研究群落组成结构和食物链关系的重要方式，但在以往的研究中，仅仅是描述性地绘出食物网有向图，很难对食物网进行定量分析。赵先生及时引进了20世纪80年代才出现的群落食物网随机理论。该理论将群落中的物种，按其在食物链中的营养位置区分为顶位物种、中位物种和基位物种；将食物链按其取食与被取食的关系区分为基-中链节、基-顶链节、中-中链节、中-顶链节四种形式，由此可以用一种全新的映射矩阵表示食物网的结构,定量分析各种链节数之间、链节数与物种数之间以及捕食者种数与猎物种数之间的关系。书中有关群落调查数据的处理、计算及分析多辅以实际例子作了详细介绍,增加了该书的实用性。赵先生结合农业生产实际,先后开展了桔园昆虫群落演替、不同种植制度菜地昆虫群落、玉米昆虫群落等有关研究。在桔园昆虫群落演替研究中,首次采用空间换时间的策略,解决了因群落演替时间过长而难于研究的困局;在柑橘群落季节变动研究中,首次利用最优分割法,在不分裂时间序列的前提下,将其划分为4个阶段,制定不同的防治措施。这些开创性的研究带动了20世纪90年代水稻、棉花、玉米、小麦、蔬菜以及多种果树等作物昆虫群落研究的热潮。

赵先生一直坚持这样一个观点,即害虫生物防治归根结底是一个生态学问题,是群落生态学理论中寄主与寄生物、捕食者与猎物相互关系的具体应用,因此在以虫治虫的生物防治中,除了对天敌昆虫的分类鉴定,更应该研究天敌昆虫的行为特点以及对害虫的寄生和捕食作用的功能反应和数量反映。赵先生以其深厚的数学功底,推导了捕食者与猎物相互关系的Lotka-Volterra方程,科学地揭示了天敌对寄主害虫在数量和时间上的跟随现象,并借以阐明了使用化学农药引起害虫再"增"猖獗的原因。

《生态学引论——害虫综合防治的理论及应用》和《群落生态学原理与方法》两本专著完美体现了赵先生深厚的数学功底与高超的语言驾驭能力。著名生态学家E.C.皮洛曾说"生态学本质上是一门数学"。虽然有些言之偏颇,但是不可否认生态学问题往往伴有复杂的数学问题,有关生态学著作中也常常因为涉及深奥的数学理论、复杂的数学公式以及烦琐的数学推导使读者望而却步。然而赵先生编著的这两本专著却让人耳目一新,诸多涉及枯燥、晦涩数学公式的生态学问题,被深入浅出地层层展现出来,数学公式因依附于具体生物学问题而变得鲜活生动。这一方面得益于赵先生过硬的数学功底和优美的语言表达能力,能够因繁就简,深入浅出,准确表达出数学模型所传递的生物学意义;另一方面则是由于赵先生具备扎实的昆虫学知识和技能,坚持学以致用、理论联系实际的原则,能够自如地将生态学的理论和方法应用于害虫控制的实践。他在著作中明确指出,对一个数学模型的探讨,我们的注意力应该放在模型的直观生物学背景、建立模型的生物学假设、各个量的生物学意义、模型应用的前提、适用的范围等方面,切忌只注意数学的技巧,而忽视对生物学基础的掌握和研究。

赵先生的两本专著在内容上有显著不同,但又相辅相成,相得益彰。正是这两本姊妹篇专著,奠定了赵先生在国内害虫生态学理论与应用研究领域的重要地位。

赵先生在昆虫生态学研究领域获得了以下重要成果奖励:"四川省农业害虫和天敌资源调查研究"获四川省科技进步二等奖(1984)、农牧渔业技术改进一等奖(1984),"柑桔叶螨种群生态系统研究"获农牧渔业部科技进步三等奖(1986);"桔园生态系昆虫群落及其控制"获农业部科技进步三等奖(1990);"柑桔园昆虫群落及害虫综合治理研究"获四川省科技进步二等奖(1991);"柑桔害螨综合管理决策支持系统研究"获四川省计算机优秀软件一等奖(1991)、科技进步二等奖(1993);"菜地昆虫群落及害虫综合防治研究"获农业部科技进步三等奖(1994);"柑桔病虫综合防治新技术研究"获农业部科技进步三等奖(1997);"角倍林主要病虫种类及角倍丰产技术研究"获重庆市科技进步三等奖(1999);"新传入检疫对象——美洲斑潜蝇的研究与防除"获重庆市科技进步二等奖(1999);《生态学引论——害虫综合防治的理论及应用》获西南农业大学重大科学成果一等奖(1985),"柑桔害虫多媒体课件"获西南农业大学优秀多媒体课件二等奖(1998)。这些成果奖励是对赵志模先生拓展和创新昆虫生态学理论,践行害虫综合防治,坚持学以致用、理论联系实际的原则的最佳注释。

2.农产品储运保护研究

农产品储运保护是植物保护工作在农作物产后阶段的延伸和继续。赵志模先生从20世纪90年代开始涉足农产品储运保护研究。作为联合国粮农组织聘请的国内专家,他全过程参与了"改进四川农户粮食产后技术"项目的研究和实施。赵先生和项目组拟定了涉及农户基本情况、粮食生产以及储粮使用、粮食储藏方式、储粮有害生物的发生、造成的损失与防除措施等10余个大项120余个小项的问答式调查提纲,走访了四川省绵阳、温江等地的320余农户,实地察看了农户储粮的种类和数量、储粮方式和用具以及有害生物发生的种类和损失。通过调查发现,农户储粮品种繁多,其储藏数量几乎达到生产量的70%。农户储粮的环境复杂，条件简陋，给害虫、害螨和储粮微生物提供了良好的滋生环境，加之农民缺乏科学的粮食保管及储粮害虫防治知识，导致粮食严重损失。在国内外专家的指导下，项目组结合农户实际，研究提出并实施了一系列简便易行、经济实用的农户储粮方法，并设计、推广了农户小型储粮仓库。该项目获得了巨大的社会、经济效益，受到时任国务院总理李鹏的赞赏，被联合国粮农组织授予"萨乌马"奖。通过该项目的实践，赵先生深感农户储粮在保证粮食安全上的重要性和迫切性。在他的提议和支持下，重庆市农业局植保植检站与西南大学植保学院合作开展了农户储粮的调查和储粮有害生物的防治工作。

在应用理论和基础理论研究方面，赵先生指导并参与了对绿豆象、四纹豆象、谷蠹、嗜卷书虱等储粮害虫的生物生态学、生化毒理学以及综合防治等的研究工作。研究发现，不同地理种群绿豆象具有明显的寄主分化现象，四川西昌种群严重危害大豆，而重庆种群主要危害绿豆和豇豆，两个不同地理种群在发育历期、存活率及繁殖力等方面都存在显著差异。四纹豆象是一种外来入侵昆虫，被列入我国出入境及国内植物检疫性有害生物名单。为了准确掌握四纹豆象的生物学特性，赵先生及其团队研究了温度和寄主对四纹豆象生物学特性的影响，比较研究了四纹豆象卵、幼虫和蛹的发育历期、存活率及成虫繁殖力等的差异，研究结果为四纹豆象的监测及综合防治提供了有力的理论依据。此外，他还参与了甲酸乙酯、高浓度CO_2气调、植物精油以及气调配合植物精油防治谷蠹、杂拟谷盗、书虱等储粮害虫的研究工作，并着力于仓储害虫应对环境胁迫适应机理的研究，综合应用昆虫生物生态学、生理生化学、环境化学、毒理学、行为学以及分子生态学等学科的知识和技术手段，为储粮害虫无公害治理提供强大的技术支撑。

随着气调储粮的推广，书虱已成为对世界粮食生产和食品安全造成严重威胁的一类害虫。针对大型国家粮食储备库和农户储粮中书虱危害日益猖獗的问题，赵先生及其团队对该虫基础生物学、与环境胁迫相关酶系的生理生化以及应用防治等方面进行了长期而系统的研究，首次报道了书虱对气调处理产生抗性的分子机理，并通过不同抗气调品系的药剂敏感性、种群适合度、内共生菌调控、呼吸代谢及主要能量代谢物质、解毒代谢酶活性差异的比较研究,阐明了书虱气调抗性与熏蒸抗性的相互关系,并进一步提出了书虱抗性进化的规律、抗气调性书虱猖獗暴发的内在机理以及抗性治理的策略。研究中提出的植物精油结合气调技术,对书虱具有强烈的忌避和熏蒸毒杀作用,为仓储书虱的绿色防控提供了新的思路。赵先生及其团队对书虱的研究,无论是在深度和广度上都处于国际领先的地位。赵先生还曾组织了两期粮食仓储保护技术国际培训班,共有20多个国家的40余储粮技术人员参加学习。在培训中赵先生主讲了粮堆生态系统的变化、调节与控制。两期培训班的举办,进一步推动了西南大学在储粮有害生物研究领域的国际合作。

赵志模先生对储藏农产品,尤其是储粮虫、螨的研究及其成果,得到了国内植保学界、储粮单位和相关研究院所的认可。他在国内外知名期刊上发表了二十余篇有关储藏物害虫研究的论文,曾被中国粮食科学院邀请作为国家重大项目"储粮有害生物防控基础研究"的论证专家,被中国农科院植保所、中国植保学会邀作《中国农作物病虫害》中储藏粮油有害生物篇撰写的主要负责人;由农业部全国高等农业院校教学指导委员会规划,经中华农业基金会立项批准,主编了全国高等农业院校教材《农产品储运保护学》。这本教材坚持理论联系实际,注意学科交叉,比较全面地反映了世界各国储运保护研究领域的新进展和我国多年来在这一领域的研究和应用成果。赵先生主编的《农产品储运保护学》并不等同于以往的储粮害虫学或农产品有害生物学,它不仅有农产品主要虫、螨、霉、鼠等有害生物形态学、生物生态学的描述,还增加了或更着重于农产品储运管理的基本环节、主要储藏方式、农产品的商品化处理和质量标准,农产品在储运期间的呼吸生理、蒸腾生理、休眠后熟与衰老陈化,粮堆生态系统的结构、功能、粮堆的理化性质及其变化与调节,农产品保鲜的基本原理与方法,与农产品储运保护有关的品质鉴定和实验研究技术等内容。赵先生把以往储粮有害生物的防控,从"术"(art)的水平提高到农产品储运保护"学"(science)的水平,这充分反映了赵先生在储运保护领域的深刻思考和学术造诣。

3. 螨类毒理学研究

我国害虫抗药性研究起步较晚,开始于20世纪50年代末、60年代初。

而相较于害虫抗药性研究,国内对螨类的研究更显薄弱。赵志模先生敏锐地捕捉到这一研究热点中的冷门,另辟蹊径,从20世纪90年代末期开始了害螨抗药性的系统研究,进而一举奠定了西南农业大学螨类毒理学与害螨抗药性研究方向在全国的领先地位。实际上,赵先生于20世纪90年代初,就与四川省植保站合作,进行有关农业有害生物抗药性的监测与治理工作。他提出并促成了省植保站"凡是进入四川境内销售的新农药,必须先进行毒力测定"的规定,以便建立该种农药在四川的敏感基线,然后持续跟踪监测该药剂在四川的抗性发展。实施这一规定在全国尚属首次。此外,赵志模先生还受聘为农业部农技推广服务中心病虫抗性对策专家组成员,受农技推广中心和四川省植保站委托,多次主持举办全国或全省的害虫抗药性及生物测定培训班,除了自编讲义和参与授课外,还为全国提供了一套包括数据转换、数学计算以及出具报告等的生物测定计算机软件。早期的这些工作为赵先生后面深入开展害螨毒理学的研究奠定了扎实的基础。

赵先生于1999年申请并获批国内第一个有关害螨抗药性的国家自然科学基金面上项目"抗药性朱砂叶螨生态适应度及抗药性风险评估研究(39970493)",该项目以棉花及多种蔬菜作物上为害严重的朱砂叶螨为研究对象,遵循种群生态学的理论和原则,综合运用昆虫生物生态学、生理生化学、农药毒理学,以及分子生物学、数量遗传学等学科知识,借助农药生物测定技术、生理生化分析技术、统计分析软件和害虫抗药性管理软件等手段,对朱砂叶螨室内种群进行系统的抗药性研究。在该项目研究中,培育获得了朱砂叶螨敏感品系,可为各种杀螨剂敏感基线的建立提供全国参考;采用不同用药策略(轮换用药、混合用药和单一连续用药)下的抗性同步筛选,比较抗性进化速率,据此评价抗性治理中不同用药措施的效果;研究了抗药性朱砂叶螨实验种群的生物学及生态学,通过王-兰-丁模型推导和适合度代价计算,发现了朱砂叶螨经阿维菌素长期汰选后形成的抗性品系反而增强了对逆境高温的适应能力的独特现象。赵先生紧跟学科前沿,老当益壮,不断将螨类毒理学及抗药性研究引向深入,2005年(时年66岁)他又申请获批了国家自然科学基金面上项目"阿维菌素抗性朱砂叶螨酯酶基因的克隆及表达研究(30571239)",该项目将西南大学,乃至全国螨类毒理学研究推进到分子生物学水平。该项研究采用RACE技术,成功地从朱砂叶螨中克隆获得两条具有完整开放阅读框的酯酶基因的 cDNA 全序列，分别命名为 *TCE1*（GenBank 登录号：EU130461）和 *TCE2*（GenBank 登录号：EU130462）；经 BLASTP 同源比对、基因结构特征基序分析和分子进化树构建表明，*TCE1* 和 *TCE2* 是两条酯酶基因，且都具有 B 型羧酸酯酶的特征结构，二者隶属于蜱螨酯酶群；基因表达与药剂抗药性关系研究表明，朱砂叶螨对杀螨剂的抗性与 *TCE1* 和 *TCE2* 的 mRNA 表达量上调有关。赵志模先生在国内外知名学术刊物上发表螨类毒理学与农螨抗药性生化及分子机理的研究论文 50 余篇，其中在国外知名刊物上发表 15 篇。

正是赵先生开创性工作所奠定的坚实基础，使得西南大学在螨类毒理学研究领域始终紧跟国际前沿。目前已在包括朱砂叶螨和柑橘全爪螨等在内的叶螨组学研究（转录组、表达谱、蛋白组和基因组）、特定基因生理毒理功能解析，以及基因表达调控通路和调控网络的鉴定等方面取得了长足进展。以"叶螨"和"杀虫剂"或"杀螨剂"为主题词检索 Web of Science 数据库发现，截至 2017 年 9 月，国内外在螨类毒理学研究领域发表 SCI 论文最多的机构是西南大学，其次是比利时根特大学，从这个意义上讲，西南大学在螨类毒理学及抗药性研究领域的能力和成果已达到国内领跑、国际并跑的水平。

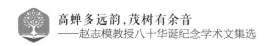

学习与工作简历

1939年4月18日	出生于四川省涪陵地区南川县(今重庆市南川区)
1944年3月—1949年9月	南川县隆化镇小学、南川县协和小学就读
1950年3月—1951年3月	南川中学农班就读
1951年3月—1953年9月	涪陵农业学校就读(其间1953年3月—9月附读农产制造专业)
1953年9月—1956年9月	万县农业学校植保专业就读
1956年9月—1960年9月	西南农学院植保专业就读
1960年9月—1962年12月	四川省凉山彝族自治州农科所就职
1962年12月—1978年10月	凉山州越西县农业局、政府农办、县委农工部就职
1978年10月—1981年10月	西南农学院植保系硕士研究生
1981年10月—1987年9月	西南农学院植保系就职
1987年9月—1988年4月	美国加州大学(UC)访问学者
1988年4月至今	西南农业大学(2005年与西南师范大学合并组建为西南大学)就职,2009年退休,科研工作和协助指导研究生至2014年

论文选集

主题一

昆虫生态学与害虫综合治理研究

昆虫寄生作用数学模式的探讨

——以纵卷叶螟绒茧蜂(*Apanteles cypris* Nixon)和稻纵卷叶螟(*Cnaphalocrocis medinalis* Guenee)自然种群为例

赵志模

Study on the Host-Parasitic Interaction with Mathematical Models-The Natural Population of *Cnaphalocrocis medinalis* Guenee and *Apanteles cypris* Nixon Used as an Example

Zhao Zhimo

Abstract: Parasite searching behavior is an important factor for determining its parasitic efficiency. Certain features of parasite searching behavior may have a marked effect on a host-parasitic interaction by three basic parasite responses: 1, the functional response to host density; 2, the response to parasitic density; 3, the response to the host distribution.

In this paper, with the natural population of *Cnaphalocrocis medinalis* Guenee and *Apanteles cypris* Nixon as an example, several models for host-parasite interaction were discussed. These models are all based on random searching where searching efficiency is either assumed to be constant (Nicholson-Bailey model) or to depend on host and/or parasitic density (Holling equation, Hassell-Varley, Hassell-Rogers model). Stability of them was analysed preliminarily also. It was not discussed for parasite searching in a non-random-the response to the host distribution.

The parameters that affect the equilibrium levels of the host and parasite population and its stability were discussed in biological control. It is concluded that a high basic searching efficiency, a high total effectual searching time, a low handling time and some degrees of interference are all optimum searching characteristics for biological control.

一、引言

应用数学模式或进行数学模拟来评价寄生昆虫的行为反应,以进一步了解昆虫寄主同寄生物之间的相互关系,对于生态学的理论研究和生物防治的实践都具有重要意义。

昆虫寄生作用的数学模式,有微分方程和差分方程两种表达形式。前者用于世代完全重叠、出生和死亡过程是连续的场合;后者则用于世代分隔清楚的离散时节(discrete time intervals)中的种群变化。本文限制在用差分方程来探讨昆虫寄生作用的数学模式。

在生态学文献中,昆虫寄生作用的数学模式很多。Royama(1971)曾对其中许多模式作过考察,它们都具有同一的基本形式:

$$N_s = N_t f[P_t, N_t] \quad \cdots\cdots(1-1)$$

$$P_{t+1} = N_t - N_s \quad \cdots\cdots(1-2)$$

式中 N_s 表示 P_t 个寄生昆虫寻找 N_t 个寄主,通过寄生作用得到 P_{t+1} 个寄生昆虫子代后的寄主残存数。在这里,所有有关寄生昆虫寻找行为的假设,都是寄主、寄生昆虫种群大小未规定的函数:$f[P_t, N_t]$。如果我们考虑最简单的情况,即寄生昆虫是单寄生和单食性的,而残存的每一个寄主的增长率相等,则昆虫寄主—寄生物种群相互变动系统的模式可概括为:

$$N_{t+1} = F N_t f[P_t, N_t] \quad \cdots\cdots(1-3)$$

$$P_{t+1} = N_t - \frac{N_{t+1}}{F} \quad \cdots\cdots(1-4)$$

式中 N_t,N_{t+1} 和 P_t,P_{t+1} 分别代表寄主和寄生昆虫相继世代的数量,F 是寄主的有效增长率(即除去寄主在世代内所有原因造成的死亡数量后实际增加的比率)。

影响函数 $f[P_t, N_t]$ 的因素很多,Holling(1966)曾以图例的形式列示了10种成分,并指出这些成分有32种可能的组合。但是,到目前为止,还没有一个模式能够包含所有这些成分或成分的组合。事实上,引入大量的成分,增多模式的变量和参数,不仅在进行数学模拟时总是比较困难,而且在实际应用上也很不方便。

Hassell & Rogers(1972)讨论了影响寄生昆虫寄生效果的三个基本反应,即:第一,寄生昆虫对寄主密度的功能反应;第二,寄生昆虫对自身密度的反应;第三,寄生昆虫对寄主分布的反应。前面两种反应都建立在寄生昆虫对寄主是随机寻找的基础上。这在数学上不仅是一个十分方便的假设,而且非随机寻找(第三种反应)的模式也是在随机寻找模式的基础上发展建立起来的。限于篇幅和资料,本文以纵卷叶螟绒茧蜂(*Apanteles cypris* Nixon)和稻纵卷叶螟(*Cnaphalocrocis medinalis* Guenee)自然种群为例,讨论上述第一和第二两种反应,并首先考虑这两种反应单独存在时的模式,然后再讨论包括了这两种反应的较为复杂的模式。

二、Nicholson-Bailey 模式

Nicholson-Bailey 模式是功能反应最基本的模式。它根据以下三个假设:第一,寄生昆虫寻找寄主是随机的;第二,一个寄生昆虫在生活期中的有效寻找面积[用"发现面积"(the area of discovery) a 表示]不变,是该种寄生昆虫固有的特征;第三,寄生昆虫有足够的卵来寄生所遇见的寄主。具有以上特征的寄生昆虫,其相遇的寄主数和寄主的密度成正比:

$$\frac{N_a}{P_t} = aN_t \quad \cdots\cdots（2-1）$$

式中,N_a 为寄生昆虫相遇的寄主数;P_t 和 N_t 分别表示寄生昆虫和寄主的密度;a 为常数,定义为寄生昆虫的发现面积。

当考虑寄生昆虫的攻击效果时,则与未被发现的寄主数成正比:

$$\frac{dN_a}{dP_t} = a(N_t - N_a)$$

积分得:

$$N_a = N_t[1 - \exp(-aP_t)] \quad \cdots\cdots（2-2）$$

由(2-1)式 $aP_t = \frac{N_a}{N_t}$

则 $N_a = N_t[1 - \exp(-\frac{N_a}{N_t})]$ $\cdots\cdots$（2-3）

由模式(2-2)可以推导出以下的亚模式和昆虫寄主—寄生物种群相互变动模式:

第一,寄生昆虫的发现面积

$$a = \frac{1}{P_t} \ln \frac{N_t}{N_t - N_a} \quad \cdots\cdots（2-4）$$

或 $a = \frac{2.3}{P_t} \lg \frac{N_t}{N_s}$ $\cdots\cdots$（2-5）

$N_s = N_t - N_a$

Nicholson-Bailey 模式假设 a 是不变的，不同的寄生昆虫受自身寻找能力的制约而有不同的 a 值。尽管这个假设不太合理，但是利用 a 值作为测度寄生昆虫效果的一个指标仍然是合适的。这不仅因为它应用于自然种群具有十分方便的优点，而且还可以对影响这个指标的因素进行分析和定量，从而有利于对寄生昆虫的效果作出评价。

例如笔者配合北碚区天敌资源调查组在歇马公社晚稻田系统调查稻纵卷叶螟幼虫密度和纵卷叶螟绒茧蜂成虫密度，并采集稻纵卷叶螟幼虫于室内饲养，计算其寄生率，按模式（2-4）计算不同时间纵卷叶螟绒茧蜂的发现面积如表1。

表1　纵卷叶螟绒茧蜂的发现面积（北碚歇马公社，1980）

日期	寄生蜂成虫密度/（头/亩）	寄主密度/（头/亩）	寄生率/%	序列	发现面积/亩	序列
Ⅶ.28	267	2 333	28.59	6	0.001 261	7
Ⅷ.9	186	1 083	36.47	3	0.002 439	4
Ⅷ.23	107	999	33.13	4	0.003 761	3
Ⅸ.4	346	999	48.05	2	0.001 893	5
Ⅸ.13	186	1 499	29.42	5	0.001 873	6
Ⅸ.25	53	833	20.29	7	0.004 278	2
Ⅹ.5	133	166	50.60	1	0.005 303	1

注：寄生蜂密度为300网（网径30公分）的数量换算值（300网面积约为0.037 5亩）；寄主密度为200丛的数量换算值（行窝距为0.6寸×0.6寸）[①]。

从表1看出，不同时间纵卷叶螟绒茧蜂对寄主的寄生率，其高低顺序和发现面积大小的顺序并不一致。这是因为前者仅仅考虑了寄主数量和被寄生数量之间的比例，而没有考虑是多少寄生蜂造成了这个比例。因此，在评价自然种群中寄生昆虫的效果，特别是比较不同种之间或同种寄生昆虫在不同条件下的寄生效果时，应用发现面积作为指标比应用寄生率作为指标，也许更为恰当。

表1所反映的纵卷叶螟绒茧蜂发现面积的差异，可能受多种因素的影响，例如当时的气候条件，寄主害虫的年龄组配等，但是作为探讨寄生昆虫

---

① 1亩≈666.7 m²；1寸≈0.033 m，1公分=1 cm。这些不是国标单位，本书为呈现原貌做保留处理，后同。——编辑注

寻找行为机制的数学模式和根据表1所提供的信息,首先应当考虑的是寄主的密度和寄生蜂自身的密度。这两个问题将在本文第三和第四部分分别讨论。

第二,昆虫寄主—寄生物种群相互变动系统模式

由(2-2)式得:

$$N_a = N_t - N_t \exp(-aP_t) \quad \cdots\cdots (2\text{-}6)$$

(2-6)式表明,被寄生的寄主数量(N_a)等于寄主的初始数量(N_t)和残存寄主之差。由此得到残存寄主数量(N_s)的亚模式:

$$N_s = N_t \exp(-aP_t) \quad \cdots\cdots (2\text{-}7)$$

将(2-7)和(1-1)式比较,显然

$$f[P_t, N_t] = \exp(-aP_t) \quad \cdots\cdots (2\text{-}8)$$

将(2-8)代入(1-3)和(1-4)则得到昆虫寄主—寄生物种群相互变动系统的模式:

$$N_{t+1} = FN_t \exp(-aP_t) \quad \cdots\cdots (2\text{-}9)$$

$$P_{t+1} = N_t [1 - \exp(-aP_t)] \quad \cdots\cdots (2\text{-}10)$$

由(2-9)式取对数

$$\lg N_{t+1} = \lg N_t - \frac{aP_t}{2.3} + \lg F \quad \cdots\cdots (2\text{-}11)$$

又由(2-5) $\frac{aP_t}{2.3} = \lg \frac{N_t}{N_s}$,

并令寄生的 K 值 $= \lg \frac{N_t}{N_s}$,则(2-11)式可写为:

$$\lg N_{t+1} = \lg N_t - K + \lg F \quad \cdots\cdots (2\text{-}12)$$

同样(2-10)式可以写为:

$$P_{t+1} = N_t - \text{anti} \lg(\lg N_t - K) \quad \cdots\cdots (2\text{-}13)$$

(2-12)和(2-13)式中的 K 值除了可以表示由寄生作用所造成的以外,还可引申为其他致死因素所引起,因此这两个模式对于应用害虫生命表来计算寄主—寄生物相互变动系,以预测寄主害虫和寄生昆虫未来发展的趋势是十分有用的。

第三,昆虫寄主—寄生物种群变动的稳定密度(steady densities)

根据(2-9)和(2-10)昆虫寄主—寄生物种群相互变动系统的模式,当

两个种群保持平衡时,应满足 $N_{t+1}=N_t=N^*$,$P_{t+1}=P_t=P^*$,因此,

由(2-9)式得出:

$F\exp(-aP^*)=1$

$$P^*=\frac{\ln F}{a} \quad\cdots\cdots\cdots\cdots\cdots\cdots\cdots\cdots\cdots\cdots\cdots\cdots\cdots\cdots\cdots\cdots(2\text{-}14)$$

由(2-10)式得出:

$P^*=N^*[1-\exp(-aP^*)]$

将(2-14)代入上式,

$\dfrac{\ln F}{a}=N^*[1-\exp(-\ln F)]$

$\dfrac{\ln F}{aN^*}=1-\dfrac{1}{F}$

$$N^*=\frac{F\ln F}{a(F-1)}\quad\cdots\cdots\cdots\cdots\cdots\cdots\cdots\cdots\cdots\cdots\cdots\cdots\cdots\cdots(2\text{-}15)$$

Nicholson-Bailey 模式只有一个平衡点(即寄主和寄生物密度恰好等于平衡密度 N^* 和 P^* 时),当寄主和寄生物密度与平衡密度稍有偏差时,就会导致两个种群增加振幅的振动,而且寄生物种群总是落后于寄主种群。虽然在极严格的实验室条件下有时观察到这种情况,但是,根据在复杂的种间关系中,波动振幅随时间而减小,故在一恒定环境内种群密度将趋于稳定的理论,这个模式是不符合的。而且在自然条件下人们也发现,昆虫寄主—寄生物相互变动系统往往是相当稳定的。然而,不能因为这样就完全否定 Nicholson-Bailey 模式,因为通过引入一个适当的作用于寄主或寄生物种群的密度制约因素,这个模式就可以趋向于稳定了。真正的问题还在于这个模式所根据的假设——随机寻找和寻找效率不变——在生物学上是否恰当。丰富的生物学资料表明,一般来说,这两个假设没有一个是真实的,也许某些寄生昆虫在生活期中的某一特定条件下是随机寻找,但绝大多数寄生昆虫却并非如此;寻找的效果也不是不变的,而是与寄主的密度,寄生昆虫自身的密度和寄主的分布有关。

三、Holling 圆盘方程

根据表1的资料,纵卷叶螟绒茧蜂的发现面积与寄主密度的关系作出

图1，表明寄生蜂发现面积的对数与寄主密度的对数呈负相关（$\lg a=1.297\,5-0.382\,4\lg N_t$，$r=0.818$，$P<0.05$），这个结论和不少生态学文献上的报道是相同的。因此，有必要进一步探讨寄生昆虫的寄生效果与寄主密度关系的机制。Holling（1959）通过模拟实验证明，寄生昆虫每当与一个寄主相遇，总是要有一定的时间用于制服、产卵或其他有关的活动，就是说，寄生昆虫在遇见一个寄主又重新寻找新的寄主之间，有一个时间间隔，这段时间叫作"处理时间" T_h（handling time）。当寄生昆虫遇到的寄主 N_a 越来越多时，其有效寻找时间 T_s 就越来越少。设寄生昆虫点寻找时间为 T，则

$$T_s=T-T_h N_a \quad \cdots\cdots（3-1）$$

根据（2-1），并令 $a=a'T_s$，则一个寄生昆虫寻找寄主的数量为 N_a，即

$$N_a=a'T_s N_t \quad \cdots\cdots（3-2）$$

式中 a' 表示寄生昆虫的攻击速率。

将（3-1）代入（3-2）则

$$N_a=a'(T-T_h N_a)N_t$$

$$N_a=\frac{a'TN_t}{1+a'T_h N_t} \quad \cdots\cdots（3-3）$$

（3-3）式就是著名的 Holling 圆盘方程。这个模式所预期的寄生昆虫对寄主密度的功能反应曲线如图2。它表明被寄生的寄主数随寄主密度的增加而上升，但当寄主密度增加到一定限度后则维持稳定状态，这就是寄生昆虫寄生数量的上限或饱和能力。

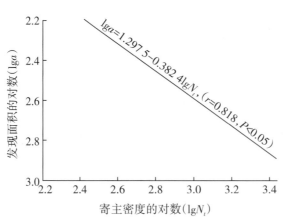

图1 纵卷叶螟绒茧蜂发现面积与寄主密度的关系
（重庆北碚，1980）

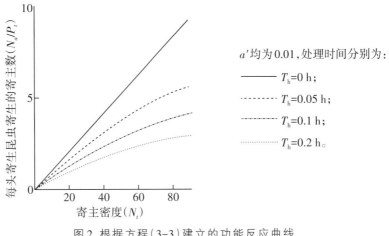

图2 根据方程(3-3)建立的功能反应曲线
(仿 Hassell & May 1973)

在(3-3)式中,当考虑被寄生寄主的总数量时,则:

$$N_a = \frac{a'TN_t}{1+a'T_h N_t} P_t$$

$$\frac{N_a}{N_t} = \frac{a'TP_t}{1+a'T_h N_t} \quad \cdots\cdots\cdots\cdots\cdots\cdots\cdots\cdots\cdots\cdots\cdots\cdots\cdots\cdots\cdots (3-4)$$

如果寻找是随机的,可以把(3-4)代入(2-3)则

$$N_a = N_t \left[1 - \exp\left(-\frac{a'TP_t}{1+a'T_h N_t}\right)\right] \cdots\cdots\cdots\cdots\cdots\cdots\cdots\cdots (3-5)$$

$$N_{t+1} = FN_t \exp\left(-\frac{a'TP_t}{1+a'T_h N_t}\right) \cdots\cdots\cdots\cdots\cdots\cdots\cdots\cdots\cdots\cdots (3-6)$$

比较 Nicholson-Bailey 模式和 Holling 圆盘方程可以看出,当处理时间 $T_h=0$ 时,Holling 圆盘方程就成了 Nicholson-Bailey 模式。这样 Holling 圆盘方程就为寄生作用提供了更好的亚模式,并显然回避了人们对 Nicholson-Bailey 模式的批评,因为该方程容许寄生昆虫的寄生效果随着寄主密度而变化,而且具有寄生数量的上限或饱和能力。它的生物学含义除了"处理时间"占去了总的寻找时间(一般来说,"处理时间"是很短的),还可以认为是寄生昆虫卵量的限制。

应用 Holling 圆盘方程来分析实验室条件下或自然条件下昆虫寄主—寄生物关系,不仅十分方便,而且具有一定的实践意义。当用观察数据拟合这个方程时,将(3-3)式作如下变换:

$$\frac{1}{N_a} = \frac{1}{a'T} \times \frac{1}{N_t} + \frac{T_h}{T}$$

令 $\frac{1}{a'T} = A$，$\frac{T_h}{T} = B$

则 $\frac{1}{N_a} = A\frac{1}{N_t} + B$ ···(3-7)

当 $N_t \to \infty$ 时，$\frac{1}{N_t} \to 0$

这时 $N_a(\max) = \frac{1}{B}$ ···(3-8)

这就是寄生昆虫的寄生上限或饱和能力。

仍以纵卷叶螟绒茧蜂寄生稻纵卷叶螟为例，并假设寄主和寄生蜂的分布是一致的，则可将表1的数据换算为每头寄生蜂寄生的寄主和平均寄主密度。其功能反应如图3，方程为：

$$\frac{1}{N_a} = 1.7303\frac{1}{N_t} + 0.1778 \quad (r=0.99, P<0.05)$$

$$N_a = \frac{0.5779 N_t}{1 + 0.1028 N_t}$$

$$N_a(\max) = \frac{1}{0.1778} \approx 6(\text{头})$$

在上式中，N_a 表示一头寄生蜂寄生的寄主数；N_t 表示相应于一头寄生蜂的平均寄主密度。如果设一定面积内寄生蜂种群的平均密度为 P，寄主种群密度为 N，被寄生的总头数为 N_A，则：

$$N_A = \frac{0.5779 NP}{P + 0.1028 N}$$

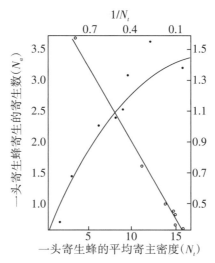

图3 纵卷叶螟绒茧蜂对寄主稻纵卷叶螟密度的功能反应(自然种群)
(重庆北碚,1980)

根据上式,就有可能将田间调查的寄主、寄生昆虫自然种群的数量直接输入方程中。

四、Hassell-Varley模式

在生态学文献中,不少作者通过实验论证了寄生昆虫的寄生效果还受到自身密度的影响,例如Hassell(1971)在研究一种寄生蜂(*Nemeritis canescens*)寄生粉斑螟(*Ephestia cautella*)幼虫的行为特点时发现,如果寄生时有两头寄生昆虫相遇,则两头或其中的一头就有离开这个寄主的趋势。当寄生蜂雌虫发现其寄主已被寄生以后,也有同样的倾向。

Hassell & Varley曾经提出了一个经验模式来描述寄生昆虫的寄生效果和自身密度之间的关系,即:

$$\lg a = \lg Q - m \lg P_t \quad \cdots\cdots(4-1)$$

$$\text{或} \quad a = Q P_t^{-m} \quad \cdots\cdots(4-2)$$

式中$\lg Q$是回归直线的截距;Q定义为当$P_t=1$时寄生昆虫的发现面积;m是回归直线的斜率,定义为寄生昆虫的相互干扰常数。

Hassell-Varley模式是描述寄生作用的一个极为重要的亚模式。它的一个主要优点在于测定其参数比较简单,例如以表1的纵卷叶螟绒茧蜂密度

和发现面积的对数作图4,它表明寄生蜂的寄生效果随着自身密度的增加而降低,其回归方程为:

$$\lg a = -0.9978 - 0.7281 \lg P_t$$

或 $a = 0.1867 P_t^{-0.7281}$

上式表明,当 $P_t=1$,即只有一个寄生蜂时,发现面积 $a=Q=0.1867$;当 $P_t>1$ 时,寄生蜂的相互干扰系数 $m=0.7281$。在评价实验室条件或自然条件下不同寄生蜂的寄生效果时,应用 Q 和 m 两个参数作为指标是具有一定实用价值的。

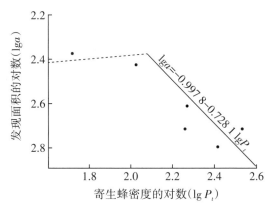

图4 纵卷叶螟绒茧蜂发现面积与自身密度的关系(自然种群)
(重庆北碚,1980)

Hassell & May(1979)认为,只有当寄生蜂密度超过某一个特定水平时,它对寄生效果的影响才是显著的。Podoler & Mendel(1979)在实验条件下研究金小蜂(*Muscidifurax raptor*)和寄主地中海实蝇[*Ceratitis capitata*(Wiedemann)]蛹之间的关系也得出了类似的结论。从图4也可以看出,当 $\lg P_t$ 小于2.2时,纵卷叶螟绒茧蜂的密度对自身的寄生效果影响不显著,而当 $\lg P_t$ 大于2.2时,其影响才是显著的。在自然条件下,由于估计寄生蜂密度上存在的误差,常常会使 m 值增高,但是除了种内的干扰外,在自然条件下还存在着种间的干扰,这样也许会起着一定的抵消作用。对于Hassell-Varley模式的评价,还在于把它引入到寄主—寄生物种群变动模式中去后,对于模式的稳定性具有重要的作用。

将(4-2)式代入(2-2)式,则:

$$N_a = N_t[1-\exp(-QP_t^{1-m})] \quad \cdots\cdots\cdots\cdots\cdots\cdots\cdots\cdots\cdots\cdots (4-3)$$

$$N_{t+1} = FN_t \exp(-QP_t^{1-m}) \quad \cdots\cdots\cdots\cdots\cdots\cdots\cdots\cdots\cdots\cdots (4-4)$$

将 Hassell-Varley 模式和 Nicholson-Bailey 模式相比较,可以看出,后面一个模式正是前面模式在 $m=0$ 时的特例。Hassell-Varley 模式在寄主有效增长率 F 和相互干扰常数 m 的广阔值域里所表现的寄主—寄生物相互变动系统都是稳定的。Hassell & May(1973)分析了模式的稳定特性,并求得稳定性界限(图5),它表明模式的稳定性完全随 F 值和 m 值而转移,而参数 a 只影响种群平衡水平,而不影响稳定性。

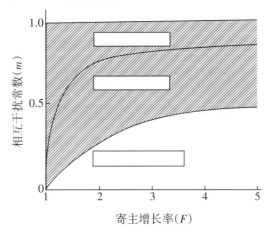

图5 相互干扰常数(m)和寄主增长率(F)间的稳定界限
(根据 Hassell-Varley 模式)
(仿 Hassell & May 1973)

五、Hassell-Rogers 模式

上述几个模式都是单独考察寄生效果与寄主密度的反应或与寄生蜂自身密度的反应。Hassell & Rogers(1972)根据 Holling 圆盘方程用一个简单的模式把寄生效果与寄主密度、寄生蜂密度两方面的反应都包括进去:

$$\frac{N_a}{P_t} = \frac{a'TCP_t^{-m}N_t}{1+a'T_h N_t} \quad \cdots\cdots\cdots\cdots\cdots\cdots\cdots\cdots\cdots\cdots (5-1)$$

式中 C 是常数,m,T,a' 与 Holling 圆盘方程和 Hassell-Varley 模式中相同的参数同义。这个模式在寄生蜂没有干扰时($m=0$)就成了 Holling 圆盘方

程,在假设没有处理时间时($T_h=0$)就成了 Hassell-Varley 模式,而在 $m=0$, $T_h=0$ 时,则成了 Nicholson-Bailey 模式。

Hassell & May(1973)以图6的三维曲面(the three dimensional surface)十分成功地表明了这个模式中,寄生效果(以 N_a/N_tP_t 表示)如何随着寄主和寄生昆虫的密度而变化。当寄主密度很高时,由于总的处理时间长(T_hN_a 值大)寄生效果不高,而在寄生昆虫密度较大时,由于相互干扰增加,寄生效果也要降低。

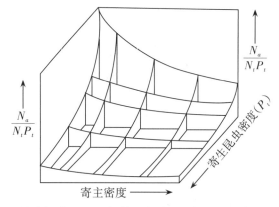

图6 以三维空间表示的寻找效果与寄主、寄生物密度的关系
(根据 Hassell-Rogers 模式)
(仿 Hassell & May 1973)

根据寄生昆虫寻找寄主是随机的假设,并令 Q' 等于(5-1)式中的常量 $a'TC$ 代入(2-3),则寄主——寄生物种群变动模式为:

$$N_a=N_t[1-\exp(-\frac{Q'P_t^{1-m}}{1+a'T_hN_t})] \cdots\cdots(5-2)$$

$$N_{t+1}=FN_t\exp(-\frac{Q'P_t^{1-m}}{1+a'T_hN_t}) \cdots\cdots(5-3)$$

Hassell & May(1973)对上述种群变动模式的稳定性进行分析后得出结论:寄生昆虫的处理时间占总寻找时间的比值总是很小的,因此对模式的稳定性作用很小。绝大多数类型的寄主—寄生物关系的稳定性,应取决于相互干扰系数 m 和寄生昆虫定向寻找(即非随机寻找)的趋势。

六、结语

寄生昆虫的行为是决定其寄生效果的一个重要因子,它通过三个基本反应影响寄主—寄生物种群的相互关系,即:(1)对寄主密度的功能反应;(2)对寄生昆虫密度的反应;(3)对寄主分布的反应。本文以 Nicholson-Bailey 模式为基础,用 Holling 圆盘方程讨论了第一种反应,用 Hassell-Varley 模式讨论了第二种反应,用 Hassell-Rogers 模式讨论了第一、二两种反应同时存在时的情况。

在生物防治中,利用寄生性天敌昆虫的成功,依赖于寄生昆虫降低寄主种群的数量并使其在一个新的低水平附近保持稳定状态。从寄主—寄生物种群的相互变动系统来考虑,这主要受两个因素的制约:一是寄主的有效增长率 F,它的大小取决于寄主的生殖力、性比以及各种因素造成的死亡率;一是被寄生的数量,它主要依赖于寄生昆虫的寻找效果及影响寻找效果的各种因素。当然,利用寄生性天敌昆虫的生物防治是否成功,还首先与它的生物学特性和生态学特点有关,例如寄生性昆虫的食性范围,对环境条件的适应能力,有无重寄生等等。但是研究寄生昆虫对寄主密度,对寄主分布的反应和行为特点,测定其几个主要参数,这对生物防治的实践,无疑是十分重要的。

在本文讨论的几个数学模式中,具有下列寻找特征的寄生昆虫,能稳定其寄主种群在一个低水平附近,从而在生物防治上较为合适(表2)。

表2 几个参数对种群稳定性和平均水平的影响

参数	在生物防治上的最适参数	对种群的影响
a		
a'	高	降低种群的平均水平
Q		
T	高	降低种群的平均水平
T_h	低	对降低稳定性极不显著,种群平均水平稍有增加
m	m 在 0—1 值域内的最适值为 $m = 1 - \left[\dfrac{F-1}{\ln F}\right]\left(2F - 1 - 2\sqrt{F(F-1)}\right)$	稳定性增加,种群平均水平有些增加

(根据 Hassell & May 1973)

描述昆虫寄主—寄生物相互变动系统的数学模式很多,国内外有不少研究、报道和综述,但多偏重于理论上的探讨和应用于控制条件下的实验种群。应当指出,把一些假设了若干限制条件的确定性模式应用于自然种群是困难的,而且也的确忽略了许多复杂的因素,但是,由于在某些情况下,被忽略的因素的确无关紧要,或者一些被忽略的因素可能会互相抵消,所以,自然种群的动态仍有可能相似于并未考虑更多复杂因素的简单模式。一般说来,从自然种群观察的数据与使用的模式拟合得越好,则模式所反映的过程越逼近于真实。

本文以田间调查的纵卷叶螟绒茧蜂和稻纵卷叶螟数据为例,主要在于说明这些模式的一般应用方法。至于文中就具体例子得出的结论,还需要通过多年积累资料,加强实验研究,进行综合分析,才能使模式日臻完善,参数更为准确。这样做,对于昆虫生态学研究的深入,对寄主—寄生物相互变动系统规律的了解以及应用于发生量的预测和生物防治的实践,将一定有所裨益。

原文刊载于《西南农学院学报》,1983年第1期

桔园昆虫群落演替初步研究

赵志模,朱文炳,郭依泉

摘要：本文应用主分量分析方法探讨了昆虫群落随树龄的演替规律,并把随树龄的演替划分为三个阶段。总的趋势是：害虫优势种从喜食嫩叶的种类如黑刺粉虱、卷叶蛾等向在老树上为害重的类群如矢尖蚧、天牛等转化,物种数由少变多,多样性和均匀度在第一阶段较低,第二阶段最高,第三阶段又下降。主分量分析还表明了引起桔园昆虫群落演替的主要昆虫类群及它们的变动规律。也比较了管理措施对桔园昆虫群落的影响,物种数以天然防治园最多,综防园次之,化防园最少,多样性、均匀度以综防园最大,其次是化防园,天然防治园最小,但群落内昆虫个体总数正好相反,天然防治园最多,其次是化防园,综防园最少。

A Preliminary Study on Insect Community Succession in Citrus Orchards

Zhao Zhimo, Zhu Wenbing, Guo Yiquan

Abstract: This paper deals with the succession of insect community of citrus orchards in Beibei, Sichuan province of China. The principle component analysis method was used for approaching the insect community succession associated with the tree ages and then three stage of succession may be divided. The main trend of succession is that the superior in sect pests change and the number of insect species increases gradually as the trees reach full growth. The diversity and evenness of the community are lower at the first stage, at the second stage they increase, but decrease again at the third stage.

This paper also analyses the influence of pest management techniques on the citrus insect community. The number of insect species expresses as following order: the orchard of natural control>integrated control>chemical control. The diversity and evenness of the community: the orchard of integrated control>chemical control>natural control. But the total number of the community: the orchard of natural>chemical control>integrated control.

在桔①园内，以柑桔为中心形成了一个复杂的生物群落，其中昆虫（包括螨类等节肢动物，下同）种类多，数量大，不仅是桔园生物群落的重要组成部分，而且极大地影响着柑桔的产量和质量。以往国内外对柑桔主要害虫和天敌进行过比较广泛的研究，但多数限于单种种群的报道。1983到1985年，我们通过室外调查和室内饲养，从群落水平上对桔园昆虫进行了研究，考察并探讨了桔园昆虫群落的组成、演替，主要类群的时间、空间，及营养生态位关系等，本文就不同树龄和不同管理条件下桔园昆虫群落的差异初步探讨桔园昆虫群落的演替规律，现报道如下。

1 材料与方法

1.1 调查方法

1.1.1 不同树龄昆虫群落调查

在重庆缙云山园艺场，选择10个不同树龄的甜橙园：Ⅰ园，苗圃；Ⅱ园，1983年定植；Ⅲ园，1982年定植；Ⅳ园，1975年定植；Ⅴ园，1973年定植；Ⅵ园，1967年定植；Ⅶ园，1960年定植；Ⅷ园，1952年定植；Ⅸ园，解放前夕建园；Ⅹ园，约1950年生。Ⅰ园为柑桔苗期，尚未定植，Ⅱ、Ⅲ园为结果初期，Ⅳ、Ⅴ、Ⅵ、Ⅶ园为结果盛期，Ⅷ、Ⅸ、Ⅹ园处于衰老更新期。各桔园均为甜橙园，除树龄差异外，其土壤、地势及管理条件基本相似。

在每个园内选有代表性的桔树5株，在每株树的东南西北四个方位上，于内外层各取一枝条，每枝记数5片叶及一尺长枝条上昆虫（包括蜘蛛和螨类）的种类和数量。天牛类，记数20株树的蛀孔数；潜叶蛾类，记数调查叶中的受害叶片数；卷叶蛾类及凤蝶类，在取样枝条附近的视野范围内记数可见种类和数量。

1.1.2 不同管理条件昆虫群落调查

在本院果园和北碚区农科所，选择三种不同管理方式的果园，即化防园、综防园和天然防治园共五个。化防园三个，其中一个柚子园、两个甜橙

---

① 为展现文章原貌，个别用字尊重历史，如"桔"（橘）不作一修改。——编辑注

园，25~35年生，常规防治，措施以喷农药为主，主要防治桔全爪螨（*Panonychus citri* McGregor）、桔始叶螨（*Eotetranychus kankitus* Ehara）、矢尖蚧（*Unaspis yanonensis* Kuwana）、桔锈瘿螨（*Phyllocoptruta oleivora* Ashmead）等，每年喷药六次左右。综防园一个，为25~30年生的甜橙园。主要措施是协调化防和生防的矛盾，控制用药次数和用量，用药方法上采取挑治以保护天敌，并在园内种植矮秆植物，为天敌提供适宜的栖息场所和补充营养，同时，加强管理，季剪除病虫枝。天然防治园为30~40年生的柚子园，30多年来，特别是近10年，基本上未经管理、害虫天敌自生自灭。在每个桔园内于1983年夏、秋、冬和1985年春各调查一次，方法是：每园选择五株，在每株树的东、南、西、北四个方位上，分别于内外层各固定一枝条，每次调查记载五片叶及一尺长枝条上昆虫的种类和数量。

1.2 分析方法

1.2.1 不同树龄桔园昆虫群落演替的主成分分析

（1）建立原始数据矩阵 Z

$$Z = \begin{pmatrix} Z_{11} & Z_{12} & \cdots & Z_{1n} \\ Z_{21} & Z_{22} & \cdots & Z_{2n} \\ \vdots & \vdots & & \vdots \\ Z_{p1} & Z_{p2} & \cdots & Z_{pn} \end{pmatrix}$$

Z_{ij}——第 j 桔园第 i 种昆虫的数量，

n——不同树龄的桔园数，

p——昆虫群落中物种数。

（2）原始数据标准化

为了消除物种间量纲不一致和数量差异的影响，对原始数据矩阵 Z，必须对属性（不同昆虫种的数量）进行标准化。本文采用离差标准化方法。标准化后的矩阵为 X，即：

$$X = \begin{pmatrix} X_{11} & X_{12} & \cdots & X_{1n} \\ X_{21} & X_{22} & \cdots & X_{2n} \\ \vdots & \vdots & & \vdots \\ X_{p1} & X_{p2} & \cdots & X_{pn} \end{pmatrix}$$

$$X = \frac{Z_{ij} - \bar{Z}_i}{\sqrt{\sum_{i=1}^{n}(Z_{ij} - \bar{Z}_i)^2}}$$

$i=1,2,\cdots,p$

$j=1,2,\cdots,n$

$\bar{Z}_i$——n个桔园第i物种的平均数量

（3）计算属性间的相关矩阵 R

因为 X 是用离差标准化处理的，所以相关矩阵 $R=XX^T$，即：

$$R = \begin{pmatrix} R_{11} & R_{12} & \cdots & R_{1p} \\ R_{21} & R_{22} & \cdots & R_{2p} \\ \vdots & \vdots & & \vdots \\ R_{p1} & R_{p2} & \cdots & R_{pp} \end{pmatrix}$$

（4）求 R 的特征根和特征向量

求特征多项式$|R-\lambda_i|=0$ 的 P 个根，并依大小顺序排列成 $\lambda_1 \geq \lambda_2 \geq \cdots \geq \lambda_p$，然后由 $uR=\lambda u$ 的关系解出 P 个 λ 相应的特征向量，并把它们依行排列就得到变换矩阵 u。其中

$$\lambda = \begin{pmatrix} \lambda_1 & & & 0 \\ & \lambda_2 & & \\ & & \ddots & \\ 0 & & & \lambda_p \end{pmatrix}$$

（5）求 N 个桔园的排序坐标

由 $Y=uX$ 可标出 N 个桔园对新坐标系的 P 个主分量坐标，但排序的目的是希望选取较少的主分量个数，在较低维空间中进行排列。一般选择两个或三个主分量作图，而忽略后面的主分量。若选择的主分量个数为 R，则 R 个主分量保留原 P 维空间的信息百分比为：

$$\sum_{i=1}^{R}\lambda_i \bigg/ \sum_{i=1}^{P}\lambda_i$$

（6）估计各物种对主分量的贡献

主分量是 P 个属性的线性组合，不能解释为单个物种的作用，我们采用负荷量估计各物种对主分量的贡献。

$$l_{ij} = \sqrt{\lambda_j} u_{ji}$$

l_{ij} 是属性 i 对第 j 个主分量的负荷量。

1.2.2 群落多样性、均匀度测定

群落多样性测定采用 Shannon-Wiener 多样性指数 H'。

$$H' = -\sum_{i=1}^{P} \frac{N_i}{N} \log_2\left(\frac{N_i}{N}\right)$$

N_i——第 i 物种的个体数

N——样本中个体总数

P——样本内物种数

H' 越大，群落多样性越高。

均匀度以实则多样性和最大多样性的比值表示：

$$J = \frac{H'}{\log_2 P}$$

J 的取值在 0~1 之间，J 越大，群落内均匀度越高。

本文所有计算均在 PG-65 型计算机上完成。

2 结果与分析

2.1 不同树龄昆虫群落的演替

2.1.1 群落演替的主分量分析

1985 年 4 月在重庆缙云山园艺场的调查结果见表 1。

表1 不同树龄桔园昆虫群落调查结果（重庆北碚，1985）

单位：个

种类	桔园									
	I	II	III	IV	V	VI	VII	VIII	IX	X
矢尖蚧	30	40	11	9	67	173	68	41	580	2 150
长牡蛎蚧	0	0	0	0	21	24	0	5	24	0
褐圆蚧	0	0	1	8	5	6	0	1	1	4

续表

种类	桔园									
	I	II	III	IV	V	VI	VII	VIII	IX	X
桔始叶螨	0	0	0	0	16	5	44	3	33	8
桔全爪螨	850	952	1 064	2 212	694	2 132	252	76	271	207
黑刺粉虱	3	3	5	64	110	86	22	62	16	25
桔蓟马	5	87	55	77	327	124	184	82	66	128
糠片蚧	0	0	2	2	4	30	0	6	6	8
卷叶蛾类	0	4	0	0	33	22	0	1	7	0
天牛类	0	0	0	0	0	0	1	4	6	6
潜叶蛾	17	15	15	4	4	2	6	2	0	1
桔蚜类	10	0	13	38	63	88	43	91	17	43
瓢虫	1	1	1	0	2	2	1	1	0	0
蜘蛛	0	2	3	1	15	0	0	1	3	2
寄生蜂	0	0	0	0	0	1	0	1	0	3

由表1计算各物种的相关系数矩阵 R,由 R 计算特征根和特征向量。表2列出了9个特征根(由大到小)及相应每轴所占信息量和累积贡献率。

表2 特征根表

序号	1	2	3	4	5	6	7	8	9
特征根	5.088 4	3.517 2	2.373 4	1.257 4	1.035 0	0.790 7	0.520 6	0.390 7	0.026 5
贡献率/%	33.92	23.45	15.82	8.38	6.90	5.27	3.47	2.60	0.17
累积贡献率/%	33.92	57.37	73.19	81.57	88.47	93.74	97.21	99.81	99.98

前三个特征根对应的特征向量及每一物种对前三个特征向量的负荷量见表3。

表3 前三个特征向量及负荷量表

物种	特征向量及负荷量					
	第一主分量		第二主分量		第三主分量	
	特征向量	负荷量	特征向量	负荷量	特征向量	负荷量
矢尖蚧	0.044 2	0.099 6	0.447 8	0.830 4	−0.089 1	−0.137 4
长牡蛎蚧	0.334 3	0.754 1	−0.026 4	−0.049 5	0.127 4	0.196 3
褐圆蚧	0.308 6	0.696 1	−0.065 4	−0.122 6	−0.312 0	−0.480 6
桔始叶螨	0.061 2	0.138 1	0.173 4	0.325 3	0.440 7	0.678 9
桔全爪螨	0.098 1	0.221 4	−0.321 5	−0.603 0	−0.432 7	−0.666 6
黑刺粉虱	0.409 2	0.923 1	−0.119 4	−0.223 8	−0.035 8	−0.055 2
桔蓟马	0.311 6	0.709 2	−0.055 0	−0.103 2	0.345 0	0.531 3
糠片蚧	0.289 9	0.656 9	0.031 5	0.059 1	−0.335 0	−0.516 0
卷叶蛾类	0.369 7	0.833 8	−0.190 8	−0.357 8	0.162 8	0.250 9
天牛类	0.023 9	0.053 8	0.495 9	0.929 4	−0.129 9	−0.002
潜叶蛾	−0.324 8	−0.732 6	−0.302 7	−0.567 8	0.015 3	0.023 6
桔蚜类	0.341 5	0.770 4	0.065 9	0.123 6	0.385 9	0.594 5
瓢虫	−0.057 7	−0.013 0	−0.300 4	−0.563 3	−0.257 4	−0.396 6
蜘蛛	0.235 0	0.530 0	−0.145 9	−0.273 6	0.075 7	0.116 6
寄生蜂	0.105 9	0.238 9	0.397 2	0.744 9	0.003 8	0.013 3

(1)从表3可以看出引起桔园昆虫群落演替的主要昆虫类群。对第一主分量,贡献最大的是黑刺粉虱,负荷量为0.923 1,其次是卷叶蛾类,负荷量为0.833 8,可以认为,第一主分量基本代表黑刺粉虱、卷叶蛾类等喜食嫩叶的种类。对第二主分量,贡献最大的是天牛类,负荷量为0.929 4,其次是矢尖蚧,负荷量为0.830 4,因而第二主分量基本上代表天牛、矢尖蚧等在老树上为害较重的类群。对第三主分量,贡献最大的是桔始叶螨,负荷量为0.678 9,其次是桔全爪螨,负荷量为−0.666 6。

(2)各物种在主分量上的符号反映了物种数量随树龄变动的趋势。这就可以提示我们进一步探讨这种差异的原因,或者是由于对环境的不同要求引起,或者是由于种间竞争关系造成。例如,在第三主分量上,两种叶螨的负荷量都比较高,桔全爪螨为-0.666 6,桔始叶螨为0.678 9,但符号相反,这反映了两种叶螨随树龄变化的不同趋势,根据对两种叶螨生物学、生态学特点的了解,其原因可能是桔全爪螨喜光,在幼树上为害重,而桔始叶螨则喜欢荫蔽环境,在较老的树上为害较重。

(3)用不同树龄桔园在第一、第二、第三主分量上的排序坐标作图1。从图中可以看出,在第一主分量上,十个桔园按树龄排列,其曲线呈凸形,这表明喜食嫩叶的害虫在柑桔树结果前期(Ⅰ园)和结果初期最少,在结果盛期(Ⅳ、Ⅴ、Ⅵ园)增多,但随着柑桔树龄的继续增加到衰老更新期(Ⅷ、Ⅸ、Ⅹ园)数量降低。在第二主分量上,随树龄增加,曲线基本上一直上升,这表明矢尖蚧、天牛等的为害随树龄增加而增加。在第三主分量上,随树龄增加其代表性物种数量变动规律不明显,这可能是因为这些物种的变化更多地受到其他条件的影响。

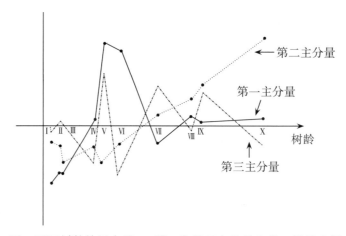

图1 不同树龄桔园在第一、第二和第三主分量上的一维排序图

(4)把10个果园在第一主分量上的取值作为横坐标,把在第二主分量上的取值作为纵坐标作图2,前两个主分量占总信息的57.37%。

从图2可以看出,桔园昆虫群落的演替趋势是:从结果前期到结果盛期,蚧类、天牛等变化不大,主要是喜食嫩叶的害虫如粉虱、蚜虫、卷叶蛾等随树龄增加而增加,从结果盛期到衰老更新期,食叶性害虫数量减少,蚧类、天牛等数量增加,到衰老更新期后,随树龄增加,蚧类、天牛等为害逐渐加重。由以上演替趋势可以将桔园昆虫群落随树龄的演替分为三个阶段：第一阶段，柑桔处于结果前期和初期（Ⅰ、Ⅱ、Ⅲ园），各种虫的数量相对较少。第二阶段，柑桔树处于结果盛期（Ⅳ、Ⅴ、Ⅵ园），主要特点是喜食嫩叶的种类如黑刺粉虱、卷叶蛾类等为害较重。第三阶段，柑桔树处于结果盛期末（Ⅶ园）到衰老更新期（Ⅷ、Ⅸ、Ⅹ园）柑桔树高大，多荫蔽，树势生长衰弱，害虫发生特点是蚧类，特别是矢尖蚧和天牛类为害较重。

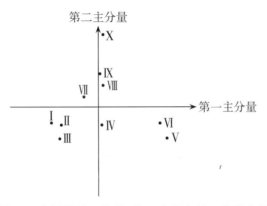

图2　10个果园第一分量、第二分量上的二维排序图

2.1.2 不同树龄桔园昆虫群落多样性、均匀度比较

桔园昆虫群落的各个演替阶段，除优势种不同外，在群落水平上必然反映出差异。现用几个主要群落指标——丰富度（调查物种数）、多样性、均匀度、样本昆虫总数研究桔园昆虫群落随树龄的变化（见表4）。

从表4可以看出，桔园昆虫群落丰富度随树龄增加而增加，多样性和均匀度在结果前期和结果初期较低，结果盛期较高，到衰老更新期后逐渐降低。群落多样性和均匀度的变动趋势一致，说明群落均匀度是多样性的主要影响因素。结果前期和初期多样性和均匀度低，一方面是由于丰富度低，另一方面，桔全爪螨数量相对较多，降低了优势度。衰老更新期后多样性较低主要原因是矢尖蚧数量很大，使均匀度降低，从而导致多样性降低。

表4 不同龄树桔园昆虫群落比较

项目	桔园									
	I	II	III	IV	V	VI	VII	VIII	IX	X
丰富度(n)	7	8	10	9	9	14	15	15	14	14
多样性(H')	0.359 4	0.724 9	0.595 8	0.577 5	2.157 6	1.275 5	2.156 1	2.604 5	1.850 5	0.984 6
均匀度(I)	0.140 9	0.241 6	0.179 4	0.182 2	0.566 7	0.326 5	0.680 2	0.666 6	0.486 0	0.258 6
样本昆虫数(N)	899	1 090	1 162	2 411	1 359	2 684	616	371	1 025	2 579

2.2 管理条件对桔园昆虫群落的影响

桔园是经常受到人为影响的农业生态系,因而害虫防治措施是另一个影响桔园昆虫群落演替的重要因素。我们仍采用几个主要群落特征指标考察害虫管理措施对昆虫群落的影响,以比较天然防治园、化防园、综防园的差异(见表5)。

表5 不同管理条件桔园昆虫群落比较

桔园	项目			
	丰富度(n)	多样性(H')	均匀度(I)	样本昆虫数(N)
天然防治园(西农柚子)	34	1.036 0	0.209 1	20 135
化防园(西农甜橙)	20	2.153 0	0.498 2	1 154
化防园(西农柚子)	24	2.264 8	0.494 0	1 982
化防园(北碚农科所)	25	2.130 9	0.458 9	6 577
综防园(北碚农科所)	33	3.054 9	0.605 6	2 294

从表5可知,由于管理措施不同,群落组成和多样性、均匀度差异很大。

2.2.1 天然防治园内丰富度和样本昆虫数都最高,但多样性和均匀度却最低

在天然防治园,由于长期未施农药,昆虫种类多,并有一些特殊种类。如桔潜叶甲在五个调查园中只在天然防治园中发现。另外,园内优势种特别突出,矢尖蚧、黑点蚧数量很大,严重时一片叶可达上千头。由此看出,在农业生态系的管理上,仅仅依靠自然防治是难于控制害虫为害的。

2.2.2 化防园内,丰富度最低,但多样性和均匀度比天然防治园高

这是由于化防园内经常施用农药防治优势种群,同时杀死大量天敌,随后害虫数量回升,有时次要害虫又可跃居为主要害虫。虽然化防园内害虫为害比天然防治园轻,但仍然不能从根本上解决害虫问题。

2.2.3 综防园内,多样性、均匀度最高,物种数也比较多,但群落内昆虫个体数却较少

因为综防园内采取了一系列措施,保护了天敌,虽然物种数较多,但优势种不突出,因而多样性、均匀度高,三种防治措施比较,综防园效果最好。

3 讨论

(1)昆虫群落的演替,是群落在时间序列上的变化。Waloff(1968)和Martin(1966)对森林的研究表明,由于单一寄主植物树龄改变,或当营养成分和树冠条件改变时,可导致昆虫群落的演替。在柑桔园内,随着树龄的增加,生育阶段的变化,植株内部营养状况和环境条件如温度、湿度、光照和土壤条件等相应地发生改变,从而导致了昆虫群落的演替。在历史资料缺少的情况下,采用一个时间断面上不同树龄、不同管理条件昆虫群落的比较,找出昆虫群落的演替规律,这虽然和历史上昆虫群落的演替有些出入,但对于预测未来昆虫群落的变化却更为实用。

(2)主分量分析研究昆虫群落演替,一方面可以找出演替的趋向,划分出演替的阶段,又可以找出引起昆虫群落演替的主要物种。我们认为,主分量分析研究昆虫群落演替是一个可行的方法。

(3)桔园昆虫群落随树龄的演替受到管理措施的极大影响,因而树龄和管理条件是决定桔园昆虫群落演替的两个主要因素,两个因素的相互作用决定了桔园昆虫群落的现状和未来的发展方向。

(4)测定群落多样性的公式最初多用于植物群落研究,昆虫群落和植物群落相比存在着显著差异。昆虫群落内各物种在营养层次上有植食性昆虫、肉食性昆虫(包括寄生性昆虫、捕食性昆虫和重寄生性昆虫等),各物种在群落内与其他物种发生联系的渠道数也不相同。例如,一个物种可以被一个、两个或多个天敌捕食(或寄生),一个天敌也可以捕食(或寄生)一个、两个或多个物种,而植物群落内各物种都处于同一营养层次,物种间只有对光照、养分、空间的竞争。作为反映群落稳定性和生产力的多样性指标,必须反映出营养层次和渠道数的差异。因此,如何应用Shannon-Wiener多样性公式研究昆虫群落还有待进一步探讨。

原文刊载于《西南农学院学报》,1985年第3期

重庆市郊不同种植制度菜地昆虫群落结构研究

赵志模,刘映红,张昌伦

摘要:文章报道了1990年6月—1991年5月对重庆市郊13种不同种植制度菜地昆虫群落的调查研究结果。调查中发现47种昆虫,按发生数量、时间和为害程度,可将害虫区分为3类。优势害虫主要是朱砂叶螨、菜缢管蚜、小菜蛾和菊潜叶蝇,各种菜地昆虫群落的多样性有所不同,但具有共同趋势,即昆虫目的多样性 $[H(O)]$ 大于目内科的多样性 $[H_O(F)]$,又大于科内种的多样性 $[H_F(S)]$;均匀度都低,优势种十分突出。模糊聚类分析表明,13种菜地昆虫群落可划分为5个亚群落,各亚群落在蔬菜种类、换茬方式以及昆虫组成上各有特点。

关键词:蔬菜;种植制度;昆虫群落

Studies on the Insect Community Structures in Vegetable Fields Adopting Different Cropping Systems in the Suburb of Chongqing City

Zhao Zhimo, Liu Yinghong, Zhang Changlun

Abstract: In 1990—1991, investigations on the insect communities in thirteen vegetable fields with different cropping system implementations were carried out in the suburb of Chongqing City. According to the occurrence amount, time and damage levels, the 47 insect species found in these investigations were divided into three parts. The dominant species were mainly carmine spider mite *Tetranychus cinnabarinus* (Boisduval), mustard aphid *Lipaphis erysimi* (Kaltenbach), diamond-back moth *Plutella xylostella* (Linnaeus) and garden pea leaf miner *Phytomyza atricornis* Meigen. Although there were different diversities among the insect communities in various vegetable fields, their tendencies were similar, i. e. the sequence of diversities were $H(O) > H_O(F) > H_F(S)$. All their evenness were low and the dominant species were highly prominent. The results of fuzzy clustering analyses showed that the insect communities in these thirteen vegetable fields were divided into five sub-communities, which were distinct each other in vegetable species, rotation systems and components of insect species.

Key words: vegetable; cropping system; insect community

菜地昆虫群落不仅因其蔬菜种类繁多而有复杂的群落结构,而且还由于蔬菜生长周期短、种收茬口不一、菜地交错、间套频繁而使得昆虫的种类和数量变化极大。作者就 1990—1991 年在重庆市郊沙坪坝区歌乐乡调查的资料,分析不同种植制度菜地昆虫群落的一些定性和定量关系,以期为蔬菜害虫综合治理提供基本信息。

调查方法

沙坪坝区歌乐乡属重庆市郊背斜低山晚菜区,海拔 450 m 左右,年均温 16.2 ℃,较沿江河谷早菜区和浅丘平坝主菜区低 2~4 ℃,是重庆市的主要淡季蔬菜基地。1990—1991 年作者在该乡选择 13 种有代表性的不同种植制度菜地(表 1),从头年大春菜栽种后的 6 月至翌年 5 月进行系统调查。每种菜地面积均在 666.7 m² 以上,随机选取 5 点,每点调查 1 m²,记载全部植株上的昆虫种类和数量。

13 种菜地的换茬时间,主要集中在 1991 年 4 月、1990 年 8 月—10 月和 1990 年 12 月—1991 年 1 月。全部菜地涉及的蔬菜包括辣椒(*Capsicum annuum* L.)、番茄(*Lycopersicon esculentum*)、茄子(*Solanum melongena* L.)、豇豆[*Vigna unguiculata* (L.) Walp]、菜豆(*Phaseolus vulgaris* L.)、莴苣(*Lactuca sativa* L.)、甘蓝(*Brassica oleracea* var. Capitata L.)、花菜(*Brassica oleracea* L. var. *botrytis* L.)、大白菜(*Brassica pekinensis*)、瓢白(*Brassica chinensis*)、萝卜(*Raphanus sativus* L.)、菠菜(*Spinacia oleracea* L.)、黄瓜(*Cucumis sativus* L.)、丝瓜(*Luffa cylindrica* Roem.)、冬瓜(*Benincasa hispida* Cogn.)、南瓜(*Cucurbita moschata* Duch.)共 16 个品种。另外,在一种种植制度的菜地中还包括一季短期红苕(*Ipomoea batatas* Lam.)。

表1 不同种植制度菜地蔬菜种类及换茬方式

菜地	1990年6月	7月	8月	9月	10月	11月	12月	1991年1月	2月	3月	4月	5月
A	丝瓜			瓢白		莴苣					南瓜	
B	豇豆		白萝卜		莴苣						豇豆	
C	南瓜		甘蓝								南瓜	
D	黄瓜		秋豇豆		莴苣+甘蓝				瓢白		黄瓜	
E	番茄		瓢白		莴苣						番茄	
F	番茄		大白菜				瓢白+花菜				南瓜	
G	辣椒		甘蓝				瓢白+莴苣				辣椒	
H	冬瓜		瓢白		莴苣						番茄	
I	黄瓜		秋豇豆		菠菜		瓢白		莴苣		黄瓜	
J	茄子		红苕		菠菜						黄瓜	
K	辣椒			莴苣							菜豆	
L	冬瓜			莴苣							黄瓜	
M	豇豆		甘蓝			莴苣					豇豆	

注：+表示间作；同一种蔬菜换茬未标出。英文字母表示不同种植制度的菜地，下同。

结果与分析

一、不同种植制度菜地昆虫群落的组成

经田间观察和室内鉴定，在13种菜地调查中共发现47种昆虫（包括螨类，下同）。其中害虫34种，分属于9个目，22个科，总个体数为52 809个；天敌13种，分属于7个目，11个科，总个体数为544个（表2）。无论是害虫还是天敌，在不同菜地中昆虫的目数、科数和种数差异不大，但昆虫的个体数，由于受不同菜地优势种群的影响，差异悬殊。

所有菜地的害虫可大体分为三种类型：①发生数量极大，为害时间长而严重的种类，包括朱砂叶螨[*Tetranychus cinnabarinus*（Boisduval）]、菜缢管

蚜[*Lipaphis erysimi*(Kaltenbach)]、小菜蛾[*Plutella xylostella*(Linnaeus)]和菊潜叶蝇(*Phytomyza atricornis* Meigen)。上述害虫在9~13种菜地发生,全年出现时间在10个月以上,调查总个体数分别为18 127、18 286、2 660和9 660头。②发生数量较大,为害时间较长而严重的种类,包括侧多食跗线螨[*Polyphagotarsonemus latus*(Banks)]、菜粉蝶(*Pieris rapae* L.)、豆野螟[*Maruca testulalis*(Geyer)]、黄曲条跳甲[*Phyllotreta striolata*(Fabricius)]、黄守瓜(*Aulacophora femoralis* Motschulsky)和筛豆龟蝽[*Megacopta cribraria*(Fabricius)]等。上述害虫在7~10种菜地发生,全年出现时间4—9个月,调查个体数从500~2 000头不等。③数量零星,为害较轻的种类,包括斜纹夜蛾[*Spodoptera litura*(Fab.)]、红腹白灯蛾[*Spilarctia subcarnea*(Walker)]、点蜂缘蝽[*Riptortus pedestris*(Fabricius)]、蟋蟀(*Gryllus testaceus* Walker)等,这些害虫一般仅在3~5种菜地发生,数量少,为害不重。

表2 不同种植制度菜地昆虫的目、科、种及个体数

单位:个

菜地	害虫				天敌			
	目数	科数	种数	个体数	目数	科数	种数	个体数
A	6	11	13	1 078	2	2	3	78
B	6	10	11	6 161	4	4	5	84
C	5	8	8	483	1	1	1	8
D	5	9	11	442	3	3	3	27
E	5	7	8	2 195	3	4	4	17
F	7	11	12	4 882	2	2	2	6
G	5	10	13	2 338	3	3	3	7
H	7	10	12	11 923	4	4	5	73
I	6	13	17	3 801	3	3	3	53
J	6	9	10	7 097	3	4	5	57
K	6	8	10	2 789	3	3	5	44
L	5	6	7	5 037	2	2	2	88
M	5	10	13	4 583	1	1	1	2
总数	9	22	34	52 809	7	11	13	544

天敌类昆虫以瓢虫最多,主要有异色瓢虫[*Harmonia axyridis*(Pallas)]、七星瓢虫(*Coccinella septempunctata* L.)、龟纹瓢虫[*Propylea japonica*(Thunberg)];蚜茧蜂类主要为菜蚜茧蜂(*Diaeretiella rapae* M'Intosh)和烟蚜茧蜂(*Aphidius gifuensis* Ashmead);绒茧蜂类主要是菜粉蝶绒茧蜂[*Apanteles glomeratus*(L.)];此外还有草蛉、蜻蜓、豆娘、虎甲及捕食螨等天敌类群。

二、不同种植制度菜地昆虫群落的特征

丰富度、多样性、均匀度和优势度是群落组织水平的几个重要的特征指标。

本文的丰富度以物种数表征,多样性采用Shannon-Wiener的信息多样性指数:

$$H=-\sum_{i=1}^{S}P_i \lg P_i \quad\cdots\cdots\cdots\cdots\cdots\cdots\cdots\cdots\cdots\cdots\cdots\cdots(1)$$

式中P_i为第i种昆虫的个体数占群落总个体数的比例,S为群落中总物种数。鉴于本调查发现的昆虫具有目、科、种的分类等级,为此群落的总体多样性用下式求得:

$$H(O,F,S)=H(O)+H_O(F)+H_F(S) \quad\cdots\cdots\cdots\cdots\cdots\cdots(2)$$

式中$H(O,F,S)$为群落总体多样性,可以证明$H(O,F,S)$等于(1)式的H;$H(O)$表示以目分类的多样性;$H_O(F)$表示目内科的多样性;$H_F(S)$表示科内种的多样性。

根据Pielou(1969)对群落均匀度的定义,本文采用实测多样性和群落最大多样性(即在给定物种下完全均匀群落的多样性)的比率表示均匀度:

$$R=H/H_{max}=-\sum_{i=1}^{S}P_i \lg P_i / \lg S \quad\cdots\cdots\cdots\cdots\cdots\cdots(3)$$

对于群落优势度的测定方法,王金福等(1988)曾采用Berger-Parker指数表示,为了使优势度含有更丰富的信息,本文提出一个新的优势度指数:

$$D=\frac{(S-1)_{max}}{N-N_{max}} \quad\cdots\cdots\cdots\cdots\cdots\cdots\cdots\cdots\cdots\cdots\cdots\cdots(4)$$

式中N_{max}为群落中个体数最多的物种的个体数,S为群落中的总的物种数,N为S个物种的总的个体数。该式的直观意义表明,优势度是指群落中

占优势的物种的个体数是其余物种平均个体数的倍数。

按以上(1)~(4)式计算,不同种植制度菜地昆虫群落的群落特征值如表3。结果表明,全部菜地调查到的昆虫(包括害虫和天敌)共47种,在不同种植制度的菜地中,虫种最多的I型菜地有20种,最少的C型和L型菜地仅有9种,结合表1可以看出,群落丰富度的高低主要受菜地蔬菜种类多少的影响。菜地昆虫群落的总体多样性亦以I型菜地最高(0.665 9),L型菜地最低(0.081 4)。各型菜地不同分类级别的多样性有一个共同的特点,即目的多样性:$[H(O)]$大于目内科的多样性$[H_O(F)]$,又大于科内种的多样性$[H_F(S)]$。假设一个极端的例子就可以明显看出这个特点的生物学背景。如果群落中的各个物种仅属于某个目中的一个科,这时目内科的多样性和科内种的多样性均等于O,并且$H(O)=H(O,F,S)$;又如果群落中昆虫的目数正好等于各自包含的科数和各科包含的种数,且各物种的个体数相等,则$H(O)=H_O(F)=H_F(S)$,并且$H(O,F,S)=H(O)+H_O(F)+H_F(S)$,由此可以看出,无论何种类型种植制度的菜地,分属于不同目的个体,在目内仅分属于较少的科,而在科内仅分属于极少的种。

群落的均匀度和优势度从不同侧面反映了各物种个体数的分布状况。一般来说,越是均匀的群落优势度越低;相反,越是不均匀的群落优势度越高。由表3看出,A、B、C、D、E、F、G、H和I等9种菜地昆虫群落的均匀度在0.365 6~0.563 3之间,其优势度为8.86~24.93;J和K两种菜地昆虫群落均匀度分别为0.306 1和0.270 2,优势度为50.53和54.27;L和M均匀度较低,分别为0.085 3和0.200 2,优势度分别达到228.99和101.62。各种菜地昆虫群落中的优势种群主要包括菜缢管蚜、朱砂叶螨、小菜蛾或菊潜叶蝇。这几种害虫都是典型的r选择的害虫。

表3 不同种植制度菜地昆虫群落特征值

菜地	丰富度 (S)	多样性				均匀度 (R)	优势度 (D)
		$H(O)$	$H_O(F)$	$H_F(S)$	$H(O,F,S)$		
A	16	0.463 7	0.122 1	0.008 4	0.594 2	0.493 5	16.58
B	16	0.456 3	0.081 3	0.001 6	0.539 2	0.447 8	24.93
C	9	0.441 2	0.028 9	0.000 9	0.471 0	0.493 6	8.86

续表

菜地	丰富度 (S)	多样性				均匀度 (R)	优势度 (D)
		$H(O)$	$H_O(F)$	$H_F(S)$	$H(O,F,S)$		
D	14	0.522 6	0.094 2	0.028 8	0.645 6	0.563 3	13.98
E	12	0.422 8	0.060 6	0.020 8	0.540 2	0.467 2	14.53
F	14	0.401 5	0.016 8	0.000 7	0.419 0	0.365 6	15.12
G	16	0.370 4	0.156 0	0.020 0	0.550 4	0.457 1	17.75
H	17	0.476 9	0.010 0	0.000 4	0.487 3	0.396 0	18.56
I	20	0.480 6	0.180 8	0.004 5	0.665 9	0.511 8	17.67
J	15	0.340 5	0.012 3	0.007 2	0.360 0	0.306 1	50.53
K	15	0.304 4	0.010 7	0.002 7	0.317 8	0.270 2	54.27
L	9	0.071 7	0.009 6	0.000 1	0.081 4	0.085 3	228.99
M	14	0.196 7	0.030 0	0.003 5	0.229 5	0.200 2	101.62

三、不同种植制度菜地昆虫群落的分类

本文采用群落系数 C 作为群落间的相似性指标：

$$C=\frac{2W}{a+b} \quad \cdots\cdots\cdots\cdots\cdots\cdots\cdots\cdots\cdots\cdots\cdots\cdots\cdots\cdots\cdots(5)$$

式中，W 为两群落共有物种中较少的个体数（如共有种个体数相等，则只用该个体数）的总和；a 和 b 分别为两群落中各物种个体数的总和，显然，当两个群落完全没有相同的物种时，$C=0$；而当两群落中的物种完全相同，且相同物种的个体数相等时，$C=1$。以 C 值建立不同菜地昆虫群落的相似性矩阵 R。该矩阵已满足反身性 $r_{ij}=1$ 和对称性 $r_{ij}=r_{ji}(i=1,2,\cdots,n)$ 的要求，为将该矩阵改造为还同时具有传递性的模糊等价关系，采用矩阵褶积：

$$R \cdot R = V(r_{ij} \bigwedge r_{ji}) \quad \cdots\cdots\cdots\cdots\cdots\cdots\cdots\cdots\cdots\cdots\cdots\cdots(6)$$

满足 $R^k = R^{2k} \triangleq R^n$ $\cdots\cdots\cdots\cdots\cdots\cdots\cdots\cdots\cdots\cdots\cdots\cdots\cdots\cdots(7)$

这时即可根据不同的 λ 值进行聚类分析。

$$R = \begin{bmatrix} 1 & & & & & & & & & & & & \\ 0.27 & 1 & & & & & & & & & & & \\ 0.43 & 0.38 & 1 & & & & & & & & & & \\ 0.53 & 0.13 & 0.28 & 1 & & & & & & & & & \\ 0.13 & 0.05 & 0.05 & 0.08 & 1 & & & & & & & & \\ 0.44 & 0.39 & 0.36 & 0.21 & 0.44 & 1 & & & & & & & \\ 0.29 & 0.58 & 0.28 & 0.14 & 0.40 & 0.67 & 1 & & & & & & \\ 0.13 & 0.19 & 0.14 & 0.06 & 0.59 & 0.37 & 0.39 & 1 & & & & & \\ 0.55 & 0.11 & 0.23 & 0.64 & 0.12 & 0.22 & 0.17 & 0.08 & 1 & & & & \\ 0.19 & 0.20 & 0.25 & 0.10 & 0.04 & 0.22 & 0.20 & 0.52 & 0.10 & 1 & & & \\ 0.49 & 0.54 & 0.45 & 0.24 & 0.13 & 0.39 & 0.70 & 0.23 & 0.28 & 0.24 & 1 & & \\ 0.53 & 0.19 & 0.31 & 0.30 & 0.37 & 0.63 & 0.52 & 0.26 & 0.34 & 0.21 & 0.42 & 1 & \\ 0.31 & 0.79 & 0.30 & 0.16 & 0.07 & 0.30 & 0.66 & 0.19 & 0.18 & 0.21 & 0.68 & 0.25 & 1 \end{bmatrix} \begin{matrix} A \\ B \\ C \\ D \\ E \\ F \\ G \\ H \\ I \\ J \\ K \\ L \\ M \end{matrix}$$

$$\quad\quad A \quad B \quad C \quad D \quad E \quad F \quad G \quad H \quad I \quad J \quad K \quad L \quad M$$

$$R^j = R^* = \begin{bmatrix} 1 & & & & & & & & & & & & \\ 0.53 & 1 & & & & & & & & & & & \\ 0.45 & 0.45 & 1 & & & & & & & & & & \\ 0.55 & 0.53 & 0.45 & 1 & & & & & & & & & \\ 0.44 & 0.44 & 0.44 & 0.44 & 1 & & & & & & & & \\ 0.53 & 0.67 & 0.45 & 0.53 & 0.44 & 1 & & & & & & & \\ 0.53 & 0.68 & 0.45 & 0.53 & 0.44 & 0.67 & 1 & & & & & & \\ 0.44 & 0.44 & 0.44 & 0.44 & 0.59 & 0.44 & 0.44 & 1 & & & & & \\ 0.55 & 0.53 & 0.45 & 0.64 & 0.44 & 0.53 & 0.53 & 0.44 & 1 & & & & \\ 0.44 & 0.44 & 0.44 & 0.44 & 0.52 & 0.44 & 0.44 & 0.52 & 0.44 & 1 & & & \\ 0.53 & 0.68 & 0.45 & 0.53 & 0.44 & 0.67 & 0.70 & 0.44 & 0.53 & 0.44 & 1 & & \\ 0.53 & 0.63 & 0.45 & 0.53 & 0.44 & 0.63 & 0.63 & 0.44 & 0.53 & 0.44 & 0.68 & 1 & \\ 0.53 & 0.79 & 0.45 & 0.53 & 0.44 & 0.67 & 0.68 & 0.44 & 0.53 & 0.44 & 0.68 & 0.63 & 1 \end{bmatrix} \begin{matrix} A \\ B \\ C \\ D \\ E \\ F \\ G \\ H \\ I \\ J \\ K \\ L \\ M \end{matrix}$$

$$\quad\quad A \quad B \quad C \quad D \quad E \quad F \quad G \quad H \quad I \quad J \quad K \quad L \quad M$$

根据矩阵 R^* 中的 r_{ij}，取不同水平的 λ 值进行聚类，其结果如图1。

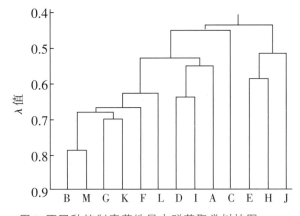

图1 不同种植制度菜地昆虫群落聚类树枝图

由图1看出,当λ取值0.55的水平时,整个菜地昆虫群落可区分为5个亚群落。结合表1可以了解各亚群落害虫组成与菜地种植制度的关系。

第1亚群落情况较为复杂,由B、M、G、K、F和L 6种菜地组成。其中B和M在λ=0.79的最高水平上并类;G和K在λ=0.70的较高水平上并类,其余F和L分别在λ为0.68和0.63的水平上与前述4种菜地并为一类。该亚群落夏菜种类较多,包括豇豆、辣椒、蕃茄和冬瓜,以朱砂叶螨为优势种群,侧多食跗线螨在辣椒上为害严重;秋菜包括甘蓝、瓢白和白萝卜,主要受菜缢管蚜为害;冬春菜基本上均是莴苣,除菜缢管蚜外,菊潜叶蝇为害亦较突出。

第2亚群落由A、D和I 3种菜地组成。其中D、I在λ=0.64的水平上并类,A在λ=0.55的水平上与D、I并为一类。该亚群落夏菜全为瓜类,主要受朱砂叶螨和守瓜类害虫为害,但优势不突出;秋冬菜包括秋豇豆、甘蓝、瓢白和莴苣,以菜缢管蚜占优势,秋豇豆上豆野螟为害比较突出。

第3亚群落由E、H两种菜地组成,两者在λ=0.59的水平上并类。该亚群落夏菜为蕃茄、冬瓜,害虫较少,秋冬菜均为莴苣,前者以小菜蛾为优势种类,后者以菜缢管蚜为害较重。

第4和第5亚群落分别为C和J两种单一的菜地。这两种菜地蔬菜种类均较单一。C型菜地仅种植南瓜和甘蓝,前者主要受守瓜类害虫为害,后者小菜蛾、菜粉蝶为害突出;J型菜地夏菜种植茄子,朱砂叶螨为优势种群;秋季种植一季秋红苕,害虫种类与蔬菜上大相径庭,冬春种植菠菜、黄瓜,害虫种类不多,但菠菜上的潜叶蝇和黄瓜上的守瓜类害虫为害比较突出。

小结与讨论

第一点:从不同种植制度入手,通过周年系统调查,研究蔬菜昆虫群落是一种新的尝试。本调查共发现47种菜地昆虫,不同种植制度菜地昆虫的种类因涉及蔬菜种类的多少,从9种到20种不等;全年调查的总个体数受优势种群的影响而变动幅度为492~11 996头。整个菜地害虫可依其发生数量的多少、为害时间的长短和为害程度的轻重区分为3类。其优势种群主要是朱砂叶螨、菜缢管蚜、菊潜叶蝇和小菜蛾;此外十字花科蔬菜上的菜

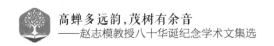

粉蝶、豆类上的豆野螟、瓜类的守瓜类害虫为害亦很突出。

第二点：不同菜地昆虫群落特征值分析表明，用Shannon-Wiener指数表示的群落多样性在0.081 4～0.665 9之间，一个共同的特点是，在总体多样性中，昆虫目的多样性大于目内科的多样性，又依次大于科内种的多样性。各菜地昆虫群落的均匀度均较低，优势度极高。其中L型菜地莴苣上的菊潜叶蝇数量达到其余昆虫平均数量的229倍。优势种十分突出是蔬菜昆虫群落普遍存在的一个重要特征。

第三点：以群落系数作为相似性指标进行模糊聚类分析，其结果表明，13种种植制度菜地昆虫群落在$\lambda=0.55$的水平上可区分为5个亚群落。各亚群落涉及的蔬菜种类、换茬方式以及相应的昆虫种类及数量均有不同。

第四点：关于不同种植制度菜地昆虫群落随时间的变化动态，拟另文报道。本文只企图为蔬菜害虫综合防治提供一些可资借鉴的信息，而无意于就此评价不同蔬菜种植制度的优劣，因为多种多样的蔬菜种类及其种植制度，对于满足人民生活需要都是重要的。

原文刊载于《植物保护学报》，1994年第1期

柑桔矢尖蚧一代幼蚧发生期数理统计预测

赵志模,吕慧平,张权炳

摘要:利用1977—1992年共16年的资料,从1—3月的23个气象因子中,通过相关分析和逐步变量选择,筛选出5个与矢尖蚧一代幼蚧发生期紧密相关的预报因子。这5个因子见2月温湿系数、3月日均温和相对湿度、1—3月日均温及大于10℃的天数。据此建立了预测矢尖蚧一代幼蚧发生期的多元回归方程、判别函数和条件频率列联表。经统计检验,对1993年的实测验证和计算历史符合率,均证明它们在理论和实践上是可行的。其中条件频率列联表预测法较为直观、简便,且可以在表中添加其后观察的数据。

关键词:柑桔矢尖蚧;多元分析;预测预报;列联表

Some Mathematical Methods for Forecasting the Occurrence Time of First Generation Larvae of Citrus Arrowhead Scale

Zhao Zhimo, Lü Huiping, Zhang Quanbing

Abstract: Based on the data from 1977 to 1992 and by applying multi-factorial and contingency table analyses, five key factors were selected from 23 meteorological elements (Jan. to Mar.), and three mathematical methods were developed for forecasting the occurrence time of 1st generation larvae of citrus arrowhead scale *Unaspis yanonensis* Kuwana. The results of 1993 forecast and historical coincidence rate in this 16 years were desirable, in which the contingency table method was most simple and easy, and capable to increase the successional data to the table.

Key words: citrus arrowhead scale; multi-factorial analysis; forecast; contingency table

矢尖蚧(*Unaspis yanonensis* Kuwana)是我国柑桔产区发生普遍而严重的害虫。该虫在四川年发生2—3代。由于越冬后的第一代发生较整齐,第二、三代世代明显重叠,加之矢尖蚧雌成蚧和2龄后雄蚧均已形成介壳,耐药性极强,所以对矢尖蚧的测报和防治都主要针对一代幼蚧或若蚧。国内迄今报道的一代幼蚧发生期预测,大多是通过系统解剖或系统调查,得出越冬雌蚧腹内卵胚眼点初见日或叶上幼蚧初见日,然后根据发育进度预测一代幼蚧高峰期和适宜的化学防治时期。这种方法虽直观、准确,但费时、费力,且雌成蚧腹内卵胚眼点及叶上幼蚧的观察较为困难。近年来,有的研究者试用数理统计方法预测,但一则由于对矢尖蚧发生的历史资料积累不足,二则进行筛选的气候因子较少,从而限制了所建方程的实际应用。本文利用重庆北碚16年(1977—1992年)矢尖蚧发生与相应气象资料,应用多元回归分析、判别分析和条件频率列联表法预测,并通过比较,以期找出回报率高、预报效果好而又较简便的统计预测方法。

1 材料与方法

(1)矢尖蚧一代幼蚧初见日资料是作者等20世纪70年代以来从事矢尖蚧研究积累起来的。各年数据通过每年3—5月、每隔5—7 d在桔园直接观察,并采集足够数量越冬雌成蚧于室内饲养(连同枝叶插入盛清水的瓶中)观察获得。

(2)气象资料由西南农业大学气象观测站(重庆北碚)提供。

(3)本文全部数据处理和计算均在微机上用STATGRAPHICS软件完成。

2 结果与分析

2.1 预报因子筛选

根据已有研究,矢尖蚧一代发生期主要受1—4月气象因子的影响。为提高预测方法的适用性,本文仅从1977—1992年各自1—3月的气象资料中提取23项超前因子,即1、2、3月各月的日均温、平均相对湿度、月降雨量、月平均温湿系数(RH/T)、月日照总时数、大于10 ℃以上的月积温(以上18项因子分别用X_{11}—X_{16}、X_{21}—X_{26}、X_{31}—X_{36}表示)、1—3月日均温、总雨量、总日照、大于10 ℃的总积温和总天数(以上5项因子分别用X_4—X_8表示)。

预报量 Y(一代幼蚧初见日)分为 4 级,各预报因子分为 5 级(表 1)。对 23 项预报因子分两步筛选:第 1 步,分别计算 23 项预报因子与预测量的单相关系数,共选出显著水平 $P \leqslant 0.05$ 的因子 10 个,它们是 X_{21}、X_{24}、X_{26}、X_{31}、X_{32}、X_{34}、X_{36}、X_4、X_7 和 X_8。因子初筛表明,1 月各项因子和各月雨量、日照与一代幼蚧初见日无显著相关。第 2 步,对初选的 10 个因子,通过逐步变量选择,逐次选入显著水平最高的因子。采用这一方法既可剔除显著性低的因子,又可以尽量保证预报因子间的独立性。此步共选出 5 个因子,它们是 2 月温湿系数 X_{24}、3 月日均温 X_{31}、3 月平均相对湿度 X_{32}、1—3 月日均温 X_4 和 1—3 月大于 10 ℃ 的总日数 X_8。

表 1　预报量(Y)和预报因子(X)的分级标准

变量	1 级	2 级	3 级	4 级	5 级
Y	4 月 10 日前	4 月 11 日—4 月 18 日	4 月 19 日—4 月 26 日	4 月 27 日以后	
X_{24}	≤7.00	7.10—8.00	8.10—9.00	9.10—10.00	≥10.10
X_{31}/℃	≤11.00	11.10—12.00	12.10—13.00	13.10—14.00	≥14.10
X_{32}/%	≤75.00	76.00—79.00	80.00—83.00	84.00—87.00	≥88.00
X_4/℃	≤9.00	9.10—10.00	10.10—11.00	11.10—12.00	≥12.10
X_8/d	≤35.00	36.00—40.00	41.00—45.00	46.00—50.00	≥51.00

2.2 多元回归预测

以 16 年矢尖蚧一代幼蚧初见日(Y)与筛选出的 5 个因子(X)建立多元回归式(与因子筛选时用逐步回归选入 5 个因子建立的回归式相同),其回归方程的方差分析如表 2,偏回归系数及其标准误、t 值和显著水平如表 3。由表 2、3 看出,该方程总体回归达极显著水平($F=12.852\ 0$,$P=0.000\ 4$),但各回归系数除常数项、X_{32} 和 X_8 的系数达显著或极显著水平外,其余均未达显著水平。从理论上讲,对未达显著水平的因子可予剔除,但为了与下述的另两种方法比较,仍保留了 5 个预报因子。该方程的历史回报率及对

1993年的实测预报见表7。

表2 回归方程方差分析

变异来源	平方和	自由度	均方	F值	P值
回归	13.629 3	5	2.725 9	12.852 0	0.000 4
剩余	2.120 7	10	0.212 1		
总计	15.750 0	15			

表3 回归方程的回归系数、标准误及t检验

变量	回归系数	标准误	t值	显著水平(P)
常量	4.724 50	0.982 98	4.806 3	0.000 7
X_{24}	−0.292 53	0.172 48	−1.696 0	0.120 7
X_{31}	0.173 58	0.154 50	1.123 5	0.287 5
X_{32}	0.413 27	0.189 77	2.177 8	0.054 4
X_4	−0.308 36	0.248 65	−1.240 2	0.243 2
X_8	−0.566 97	0.221 63	−2.558 2	0.028 5

2.3 判别分析预测

当预报量区分为两级或多级时,可以从表征各级的预报因子量(预报因子可以是分级或不分级的数量数据,为了进行不同预测方法的比较,本文均采用同一标准区分的分级数据)建立P个函数(P=预报量级数−1),取特征值最高的1个函数(一维判别)或最高和次高的两个函数(二维判别),可以判别一组预报因子可能属于预报量的哪一级。判别函数同多元回归方程一样有如下形式:

$$\hat{Y}=a+b_1X_1+b_2X_2+\cdots+b_nX_n$$

式中a为常量,b_i为各预报因子的系数,X_i为非标准化的预报因子数据。判别时,将各预报因子的值代入判别函数,根据计算的$\hat{Y}$值与预报量各级形心(centroids)的亲合程度确定预报级别。

表4列出了由16年资料导出的3个判别函数的特征值、相对贡献率和

典型相关系数。由表4看出,第1个函数的特征值远比第2、3个函数的高,它已能反映预报量90.58%的信息,其典型相关系数为0.931 39。因此本文只选用第1个函数(采用一维判别可以避免二维、三维判别时计算复杂或需作图的麻烦)。该函数的各项系数及预报量各级形心值如表5。由表5的预报量各级形心值看出,这些值没有重叠的现象,说明该函数能很好地判别出预报量的级别。为了确定计算值$\hat{Y}$与各级形心的亲合度,取相邻级别形心值的中值作为预报级别的界限,看$\hat{Y}$值落入哪一预报级别的范围。1级:$\hat{Y} \geq 3.688\ 7$;2级:$1.593\ 3 \leq \hat{Y} < 3.688\ 7$;3级:$-0.826\ 3 \leq \hat{Y} < 1.593\ 3$;4级:$\hat{Y} < -0.826\ 3$。该判别函数的历史回报率及对1993年的实测预报结果如表7。

表4 导出的3个函数的特征值、相对贡献率及典型相关系数

判别函数	特征值	相对贡献率/%	典型相关系数
1	6.546 58	90.58	0.931 39
2	0.533 02	7.38	0.589 65
3	0.147 68	2.04	0.358 72

表5 第1判别函数各变量的系数及预报量的各级形心值

变量	变量的系数	预报量级别	级形心值
X_{24}	0.780 61	1	4.551 31
X_{31}	−0.407 94	2	2.826 07
X_{32}	−1.021 20	3	0.360 46
X_4	0.721 89	4	−2.013 06
X_8	1.514 01	—	—
常量	−4.377 24	—	—

2.4 条件频率列联表预测法

将预报量Y与5个预报因子的条件频率列联表如表6。预报由该表综合5个预报因子计算X_j出现k级时,Y出现P级的平均条件频率$\hat{Y}_P$($\hat{Y}_P = \frac{1}{5}\sum_1^5 P_{j/k}$),取最大的$\hat{Y}_P$所对应的级别作为预报级别。

例如1977年X_{24}、X_{31}均为4级,X_{32}、X_4、X_8均为1级,则:

$$\hat{Y}_1=\frac{1}{5}(0+0+\frac{2}{5}+0+0)=0.08$$

$$\hat{Y}_2=\frac{1}{5}(0+0+0+0+0)=0$$

$$\hat{Y}_3=\frac{1}{5}(\frac{1}{5}+\frac{4}{5}+\frac{2}{5}+\frac{1}{2}+\frac{1}{7})=0.4086$$

$$\hat{Y}_4=\frac{1}{5}(\frac{4}{5}+\frac{1}{5}+\frac{1}{5}+\frac{1}{2}+\frac{6}{7})=0.5114$$

其中以$\hat{Y}_4$最大,预报为4级,与实际发生情况相符。

表6 预报量Y与5个预报因子的频率列联表

项目		$\hat{Y}$的级别				项目		$\hat{Y}$的级别			
		1	2	3	4			1	2	3	4
X_{24}的级别	1	1/2	0	1/2	0	X_{32}的级别	4	0	0	0	1/1
	2	1/2	0	1/2	0		5	0	0	0	0
	3	0	1/4	1/4	2/4	X_4的级别	1	0	0	1/2	1/2
	4	0	0	1/5	4/5		2	0	0	2/7	5/7
	5	0	0	2/3	1/3		3	0	0	3/4	1/4
X_{31}的级别	1	0	0	0	2/2		4	1/2	1/2	0	0
	2	0	0	0	2/2		5	1/1	0	0	0
	3	0	0	1/2	1/2	X_8的级别	1	0	0	1/7	6/7
	4	0	0	4/5	1/5		2	0	0	3/4	1/4
	5	2/5	1/5	1/5	1/5		3	0	0	0	0
X_{32}的级别	1	2/5	0	2/5	1/5		4	0	1/3	2/3	0
	2	0	1/6	4/6	1/6		5	2/2	0	0	0
	3	0	0	0	4/4						

表7 三种预测方法的回报率及对1993年的实测预报

年份	观察值 Y	多元回归法		判断分析法		列联表法	
		$\hat{Y}$	级别	$\hat{Y}$	级别	$\hat{Y}_p$的最大值	级别
1977	4	3.786 7	4	-1.671 9	4	0.511 4	4
1978	3	3.069 1	3	0.097 4	3	0.470 0	3
1979	3	2.759 9	3	0.950 9	3	0.676 8	3
1980	3	2.618 8	3	1.344 7	3	0.580 6	3
1981	1	1.352 4	1	4.580 7	1	0.560 0	1
1982	3	2.467 4	2*	1.731 5	2*	0.676 8	2
1983	4	3.837 0	4	-1.935 9	4	0.647 6	4
1984	3	3.907 4	4*	-1.912 5	4*	0.555 4	3
1985	4	4.489 9	5	-3.570 4	4	0.814 2	4
1986	3	3.151 1	3	-0.049 2	3	0.486 2	4
1987	1	1.336 6	1	4.522 0	1	0.660 0	1
1988	4	3.784 1	4	-1.768 6	4	0.874 2	4
1989	4	3.838 8	1	-1.803 8	4	0.680 8	4
1990	4	3.601 3	4	-1.164 4	4	0.500 0	4
1991	2	2.040 1	2	2.826 1	2	0.356 8	3*
1992	4	3.957 7	4	-2.176 5	4	0.874 2	4
1993	3	3.447 6	3	-0.803 6	3	0.410 0	3
回报率		87.50%		87.50%		93.75%	

注：*表示回报与实际观察值不符；1993年资料未进入统计计算，用以检查预报效果。

3 小结与讨论

第一，矢尖蚧一代幼蚧发生期预测主要是为化学防治适期提供依据。已有研究表明，矢尖蚧一代幼蚧初见后有两个明显的高峰期。因此选定叶上幼蚧初见日后的21、35和49 d，即一代幼蚧的双峰期和终见期作为矢尖蚧化学防治的关键时期。

第二,数理统计测报中,衡量一个预测式可靠与否的标准是自身的统计检验和实测验证,但历史回报率的高低也是一个重要参考。本文建立的多元回归方程和判别函数都达极显著标准;三种预测方法对1993年的实测与实际发生情况相符;其历史回报率,条件频率列联表法为93.75%,其余两种方法为87.50%。因此从理论和实践上看,这三种方法都是可行的。值得提出的是,预报量级别的划分标准不是一个特定的时刻,而是一个时段,因此由多元回归方程和判别函数计算的结果,可以确定为一代幼蚧初见日的大体时间段,而无须拘泥于某个固定的级别。就条件频率列联表来说,有时也会出现两个级别的 Y 值相近或相等的情况,这时亦可用上述方法灵活处理。

第三,多元回归方程和判别函数都具有同一的 $\hat{Y}=a+b_1X_1+b_2X_2+\cdots+b_nX_n$ 的线性形式,但前者计算的 Y 值即为预报级别,而后者计算的 Y 值还需与预报量各级的形心值相比较才能确定预报级别。以本文的结果而言,笔者认为采用多元回归预测法较判别函数法简便。条件频率列联表法不仅直观性强、计算简便,而且在已建列联表的基础上可以把以后逐年的实际观测资料添加进去,其他两种方法如要添加资料必须重新建立方程或函数,在缺乏计算机的情况下,这终究是比较困难的。

第四,当以分级数据建立预测式或列联表时,级别划分的标准十分关键。通常,划分级别要考虑它的直观生物学意义,要操作方便,据此建立的方程能通过统计检验。因此,确定级别常需作多次尝试,实际上这也是逐步探索预报量与预报因子相互关系的过程。

原文刊载于《植物保护学报》,1995年第3期

舞草种子的蚂蚁传播

张智英,曹敏,杨效东,赵志模

摘要:舞草[*Codoriocalyx motorius*(Houtt.)Ohashi]由于其小叶具有自身"摆动"的功能,从而具有较高的观赏价值。在长期的演化过程中,舞草与蚂蚁形成了互惠共生的关系:舞草种子生成了附生其上的能吸引蚂蚁的油质体,蚂蚁在搬运取食中,使舞草种子得以传播,舞草种子最重要的传播者是圆叶铺道蚁(*Tetramorium cyclolobium* Xu et Zheng)和布立毛蚁(*Paratrechina bourbonica* Forel)。另外长足光捷蚁(*Anoplolepis gracilipes* Smith)和两种大头蚁(*Pheidole* sp.1 和 sp.2)也搬运其种子。野外试验表明,圆叶铺道蚁日搬运活动与气温呈显著正相关,即 Y(搬运种子数)= $-9.5038+0.5608X$(气温)($r=0.7196**$, $n=33$, $P<0.01$),中午搬运效率达到高峰。布立毛蚁日搬运活动在上、下午各有一个高峰,上午的高峰出现时间不稳定,下午的高峰出现在 16:00—18:00。舞草种子上附生的油质体是吸引蚂蚁并产生搬运行为的主要物质。化学分析表明,油质体富含蚂蚁生长发育所必需的 10 种氨基酸和多种无机元素。样地采用陷阱诱捕蚂蚁的调查显示,5 种传播者中,圆叶铺道蚁数量最大,分别占蚂蚁总量的 8.26% 和搬运蚂蚁总量的 48%。这说明圆叶铺道蚁在舞草种子的搬运中起着主要作用。

关键词:舞草;种子;蚂蚁;搬运

Codoriocalyx Motorius Seed Removed by Ants in Xishuangbanna

Zhang Zhiying, Cao Min, Yang Xiaodong, Zhao Zhimo

Abstract: Many relationships between plants and insects have formed during a long evolution, and mutualism is one of these relationships and therefore is a trend of coordinating evolution between them. Many reports on seed dispersal by ants have been given out outside of China at present, it is still attracting many researchers, and the focus is about removal rate of seed, that is RRS, as well as affecting factors. In China, reports on the topic are very few, and the study on mutualism between ants and *Codoriocalyx motorius* (Houtt.) Ohashi is not reported oversea yet. The study has found out that *Codoriocalyx motorius* is a myrmecochore and several species of ant remove seed of *Codoriocalyx motorius* as well with phenomena of the mutualism between them.

The study was made in Xishuangbanna district, a tropic area, Yunnan Province of China. The geographical location of the spot is 21°41′ N and 101°25′ E, and the elevation is 600 m above sea level. The annual mean temperature is 21.4 ℃ and annual mean rainfall is 1 557 mm with relative humidity being 86%. The time of every year is divided into two periods: rain period (Jun. to Oct.) and dry period (Nov. to May).

Codoriocalyx motorius is a small perennial shrub of Leguminosae. One of its values that attract people is of self-movement of its leaves obeyed to audio frequency around it. Its seed is average 3.18 mm in length, 2.58 mm in width and 6 985 mg in weight. An opal elaiosome inter grows on the hilum, and therefore attracts ants for removal.

Authors have found out five ant species that remove seeds of *Codoriocalyx motorius*. They are *Tetramorium cyclolobium* Xu et Zheng, *Paratrechina bourbonica* Forel, *Anoplolepis gracilipes* Smith and two *Pheidoles* (*P.* sp. 1 and *P.* sp. 2). In the test spot, by pitfall trapping way 436 ants are collected, among which there are 36 *Tetramorium cyclolobium*, making up of 8.26% of the total, 8 *Paratrechina bourbonica* and 1.83%, 23 *Anoplolepis gracilipes* and 5.27%, and 8 *Pheidole* sp. 1 and 1.83% *Pheidole* sp. 2 is not found out in the trapper. The test result shows *Tetramorium cyclolobium* plays a greater role in removing seeds of *Codoriocalyx motorius*.

The above 5 ants were observed for investigating their activity of removing the seeds of *Codoriocalyx motorius*.

Workers of *Tetramorium cyclolobium* look for food alone. When finding out food, it leads its fellows to move the food. Although being small, the worker of *Tetramorium cyclolobium* can alone remove one seed. During removing seeds, these ants often stop for searching out road and then go back to carry seeds. The daily intensity of removing seeds for *Tetramorium cyclolobium* increases with rising of temperature, and generally reaches a peak at noon when daily temperature goes to the highest point. Total number of removed seeds for *Tetramorium cyclolobium* is highly positive relative to daily temperature, and the linear regression equation between them is:

$Y = -9.503\ 8 + 0.560\ 8X$ (Y: total number of removed seeds; X: temperature ℃)

($r = 0.719\ 6^{**}$, $n = 33$, $P < 0.01$)

The results from field tests show that *Paratrechina bourbonica* has two peaks of removing seeds every day, one being from 10 am to 12 am, with not good stability, and another being from 4 pm to 6 pm with good stability. When temperature exceeds 25 ℃ at noon, activity of removing seeds reduces. Total number of removed seeds by *Paratrechina bourbonica* is not significantly relative to daily temperature.

When there are enough food and there are no other ants coming to get food, *Tetramorium cyclolobium* and *Paratrechina bourbonica* do not directly move seeds.

And therefore, these two species of ants may bite elaiosomes bit by bit and take them to their nest. It is very obvious for them to do so in order to save their energy. However, when they compete for food and their numbers are near equal, *Paratrechina bourbonica* will drive *Tetramorium cyclolobium* away. But *Tetramorium cyclolobium* will expel *Paratrechina bourbonica* too when its number is over the number of *Paratrechina bourbonica* and sees the latter to act alone.

Anoplolepis gracilipes looks for food by single too, but it is not seen that they queue for moving seeds together when they find out seeds. Its moving distance is longer and may covers 7—8 m. *Anoplolepis gracilipes* still loots seeds from other species of ants that are moving seeds.

Pheidole sp. 2 moves one seed by single and will still transport the seed whose elaiosome is off. *Pheidole* sp. 1 also looks for food by single, but when seeds are found out, the fellows of *Pheidole* sp. 1 can come forward to move. Generally, one *Pheidole* sp. 1 moves one seed. And however, when the seed is moved near to the nest, other fellows can help for moving the seed into the nest. *Pheidole* sp. 1 is the smallest, the mean weight being 0.000 138 g, and however the weight of one seed is 0.006 98 g, being 50.6 times of *Pheidole* sp. 1's weight.

Investigation of the nests shows that *Tetramorium cyclolobium* and *Paratrechina bourbonica* only eat elaiosome of *Codoriocalyx motorius* and leave the complete seeds in the nests. *Tetramorium cyclolobium* and *Paratrechina bourbonica* are the most important removers among these ants.

The experiments that elaiosomes attract ants give out the results that the 80%—90% of seeds retaining elaiosome are removed away by ants within 30 minutes after seeds are laid out, and those seeds with no elaiosome are left, and all of seeds retaining elaiosome are moved within 60 minutes and therefore 10%—20% of seeds without elaiosome are also removed. The observation also finds out that the ant, which removes seeds without elaiosome, is mainly *Pheidole* sp. 2 and others ants do not move those seeds retaining no elaiosome. Therefore, the activity that ants remove seeds of *Codoriocalyx motorius* is mainly originated from fact that seed's elaiosome attracts ants.

According to analyzed results, the seed elaiosome contains 17.85% of protein, 13.1% of amino acids. There are 18 kinds of amino acids, 10 kinds of which are basic for ants, respectively being arginine, histidinol, leucine, isoleucine, lysine, methionine, alanine, threonine, tryptophan and valine. Contents of arginine and histidinol as well with aminoglutaric acid and proline make respectively up more than 10% of total amino acids.

Seed elaiosome also has 11 kinds of inorganic elements, 8 kinds of which are necessary for ants: respectively being calcium, phosphorus, iron, potassium, copper, manganese, magnesium, zinc.

Key words: *Codoriocalyx motorius*; seed dispersal; ants

植物与昆虫在长期的演化过程中形成了许多复杂的相互关系,互惠共生是其中的一种,并且是植物与昆虫协同演化的趋势。互惠共生指两个种群的相互作用对各自种群的增长都有促进作用。这种现象在自然群落,特别是在热带生境中普遍存在,并对维持生态系统的平衡起着重要作用。蚂蚁与植物的互惠共生关系表现为前者依赖后者获得营养、能量和适宜的栖境,而后者则在免遭其他害虫为害、传播其种子和偶尔为其传粉等方面获得收益。蚂蚁作为植物种子的传播者和授粉者在生态和进化上具有重要意义,近十几年有关的研究主要集中在这两个方面。蚂蚁搬运传播种子的现象比较普遍,至少在83科植物上被发现。通常人们把依靠蚁类携带散布种子的植物称为蚁运植物。这类植物的种子常附生有富含蛋白质、脂肪和油类等蚁类嗜食成分的种阜,它能作为诱饵吸引蚁类并借以完成自身的散播,而带有油质体的植物种子特别能吸引蚂蚁。

蚂蚁传播植物种子的研究国外开展得较多,且目前仍是一个活跃的研究领域。研究的内容主要集中在蚂蚁传播种子的效率(Removal Rate of Seed,简称RRS)及其影响因子上。目前国内鲜见这方面的报道,而有关舞草种子与蚂蚁互惠共生现象的研究在国外亦未见报道。

本文观察了几种蚂蚁对舞草种子的搬运传播行为,并就搬运种子的效率、种子油质体化学成分及对蚂蚁的引诱作用等进行了初步研究。

1 材料与方法

1.1 样地

研究地点位于西双版纳勐腊县勐仑镇热带植物园,地处于北纬21°41′,东经101°25′,年均温21.4 ℃,年降雨量1 557.0 mm,相对湿度86%,有雨季、旱季之分,雨季在每年的6—10月份,其余月份为旱季。试验选择在植物园苗圃地进行,内有几十种引种植物,其中舞草是1999年4月从思茅孟连引进栽培的。

舞草[*Codoriocalyx motorius*(Houtt.)Ohashi]属多年生豆科小灌木,树高约1 m,初生真叶为一对生小叶,此后为互生单叶,8叶龄后多为互生三出复叶,中出大叶长椭圆形或披针形,侧出小叶为倒披针形,大叶的长度和宽度分别为小叶长和宽的5～7倍和4～5倍。舞草种子于每年10—12月份成熟。荚果镰形或直,长2.5～4.0 cm,阔约5.0 mm,腹缝线直,背缝线稍缢缩,成熟时沿背缝线开裂,裂口朝下,种子掉落地上。每个荚果可产5～9粒种子。种子深褐色,表面具蜡质,种脐处附生乳白色的油质体,易与种子剥离。种子长平均3.18 mm,宽2.58 mm,千粒重6.985 g,油质体约占种子总质量的3.6%。

1.2 方法

1.2.1 蚂蚁搬运舞草种子行为的观察

将新鲜成熟的舞草种子随机放在样地里,观察记录蚂蚁搬运情况,当蚂蚁将其搬运15 cm以上距离后,分种采集部分蚂蚁装于盛有75%乙醇的小瓶中,带回室内鉴定。

1.2.2 主要蚂蚁种类日搬运活动及搬运种子效率比较

将新鲜成熟的种子15粒随机放在舞草植株茎杆周围10 cm的土壤中,放置环境与种子掉入土壤的自然状态一致。从8:00—19:00,每小时记录一次温度和各种蚂蚁搬运的种子数,并随时补充种子,连续观察记录3 d。

1.2.3 种子油质体对蚂蚁的吸引作用

将10粒除去油质体的种子与10粒未除去油质体的种子放在地上,间隔环状排列,在30 min和60 min时,分别观察记录种子被蚂蚁移动的数量,重复3次。

1.2.4 样地蚂蚁调查

采用陷阱诱捕法。将口径为7.5 cm的一次性塑料杯埋入地下,杯口与地面齐平,杯四周用泥土填平,杯中盛1/3的3%乙醛溶液(防止蚂蚁逃跑和腐烂)。样地共设16个陷阱,按对角线放射状排列,陷阱间相隔2 m。放置48 h后,将诱集到的蚂蚁移入盛有75%乙醇的小瓶内分别标记保存,带回室内进行种类鉴定和数量统计。

1.2.5 种子油质体营养成分分析

种子采自思茅地区景谷县野外舞草大片生长地,室内人工取其油质体分析。蛋白质采用凯氏定氮法;氨基酸采用盐酸水解法,用 HITACHI 835-50 测量;色氨酸采用比色法;无机元素采用 ICP-AES 法。

2 结果与分析

2.1 搬运舞草种子的蚂蚁种类及其搬运行为

本试验共观察到 5 种蚂蚁搬运舞草种子,它们是圆叶铺道蚁(*Tetramorium cyclolobium* Xu et Zheng)、布立毛蚁(*Pharatrechina bourbonica* Forel)、长足光捷蚁(*Anoplolepis gracilipes* Smith)和两种大头蚁(*Pheidole*, sp. 1 和 sp. 2)。圆叶铺道蚁工蚁单独搜寻食物,当发现食物后便召集同伴列队前去搬运。尽管该蚁体小,但通常是一个蚂蚁搬运一粒种子,少数是 2 只蚂蚁合搬一粒。它们搬运的速度较慢,搬运 1 m 的距离约需 30 min。在搬运过程中,常将种子放下,四处探路,然后再回头将种子搬走。早晚均可见到活动,放在样地各处的种子,很快便被该蚁搬走。布立毛蚁在上午和下午各有一个活动高峰,工蚁也是单独搜寻食物,发现食物后列队搬运,通常一只蚂蚁搬运一粒种子。它们搬运的速度较快,并且选择最近的路线到达蚁巢。有时该蚁会将种子搬运到途中的洞穴里,将油质体分离后再快速运回。在食物多又没有其他蚁群竞争时,圆叶铺道蚁和布立毛蚁可不直接搬运种子,而是将种子上的油质体一点一点咬下搬走,显然这可以节省能量。但是通常在寻找和搬运途中都存在着种间竞争。在争夺种子资源时,如果圆叶铺道蚁的数量与布立毛蚁相差不大,一般布立毛蚁会将圆叶铺道蚁赶跑;而当圆叶铺道蚁的数量多而布立毛蚁只是单个活动时,则后者会被前者赶离种子。长足光捷蚁也是单个寻找食物,发现食物后不见列队搬运的现象,它搬运的距离较远,可达 7~8 m,搬运的速度也最快,搬运时种子不离身,会上树。通常它会在其他几种蚂蚁搬运途中将搬运者吓跑,然后将种子搬走。大头蚁 sp. 2 在搬运种子时一般也是一头蚂蚁搬运一粒种子,它在仅剩下没有油质体的种子时也会把种子搬走。大头蚁 sp. 1 也是单个搜寻食物,当发现种子时有同伴前来搬运,一般一只蚂蚁搬运一粒种子,当快搬运到蚁巢时,许多头蚂蚁会合力搬运一粒种子。为了腾出空间贮存种子,该蚁会将蚁巢里面原先储放的昆虫肢体如翅、足等清理出来。大头蚁 sp. 1个体最小,据测定每头平均重0.000 138 g,而一粒舞草种子平均重0.006 985 g,种子的质量是该蚂蚁的50.6倍。跟踪蚁巢调查显示,圆叶铺道蚁和布立毛蚁只取食种子油质体而留下完好无损的种子。

2.2 主要蚂蚁种类日搬运活动及搬运种子效率比较

搬运舞草种子的蚂蚁主要是圆叶铺道蚁和布立毛蚁两种。这两种蚂蚁日搬运活动见图1,搬运种子效率见图2。

从图1看出,布立毛蚁日搬运有两个高峰,一个高峰出现在上午的10:00—12:00,另一个在下午的4:00—6:00,中午气温超过25 ℃,活动减少。上午高峰出现的早晚受当日日出时间和温度的影响较大。17日日出较早,温度较高,高峰出现的时间为10:30;18日日出稍后,且温度低于17日,高峰出现的时间为11:00;而19日日出在12:05,因此,上午搬运活动很少,未出现高峰。下午搬运高峰比较稳定,3 d观察都在16:00—18:00。圆叶铺道蚁日搬运活动随温度升高而加强,一般在中午温度最高时达到高峰,搬运种子量与日气温呈正相关,即$Y=-9.503\ 8+0.560\ 8X$,(Y:搬运种子数,X:气温℃;$r=0.719\ 6^{**}$,$n=33$,$P<0.01$)。布立毛蚁搬运量与日气温相关性不显著。

图2显示,布立毛蚁的日搬运效率比圆叶铺道蚁高,这是由于布立毛蚁个体较大,搬运的速度比圆叶铺道蚁快,在竞争中处于优势。

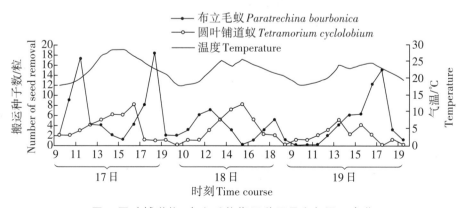

图1 圆叶铺道蚁、布立毛蚁搬运种子量和气温日变化

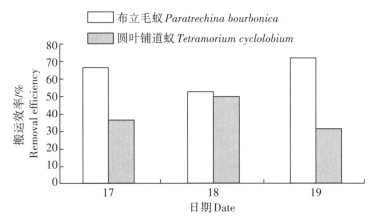

图2 圆叶铺道蚁、布立毛蚁搬运种子效率比较

2.3 种子油质体对蚂蚁的吸引作用

舞草种子放置30 min后,即有80%~90%的带油质体的种子被搬走,留下所有无油质体的种子;60 min后,所有带油质体的种子均被搬走,而且有10%~20%的无油质体的种子也被搬走。据观察,无油质体的种子主要是由大头蚁sp.2搬走的,没有发现其他几种蚂蚁搬运无油质体种子的现象。由此可见,舞草种子主要以其油质体吸引蚂蚁搬运。

2.4 搬运舞草种子的蚂蚁在样地内所占的比重

用陷阱诱捕法在样地内共获蚂蚁436头。其中圆叶铺道蚁36头,占样地内蚂蚁总量的8.26%;布立毛蚁8头,占1.83%;长足光捷蚁23头,占5.28%;大头蚁sp.1有8头,占1.83%;样地陷阱内未发现大头蚁sp.2。由此也可看出,圆叶铺道蚁在舞草种子的搬运传播中起着主要作用。

2.5 种子油质体营养成分

种子油质体含蛋白质17.85%,氨基酸总量13.12%,氨基酸种类及含量见表1,无机元素种类及含量见表2。舞草种子油质体含有的氨基酸种类较多,在已测定的18种中:以脯氨酸、谷氨酸、组氨酸、精氨酸含量较高,均占总量的10.00%以上;而蛋氨酸、缬氨酸的含量较低,分别占总量的0.15%、

0.76%。种子油质体包含了昆虫营养上所必需的10种氨基酸,即:精氨酸、组氨酸、亮氨酸、异亮氨酸、赖氨酸、蛋氨酸、丙氨酸、苏氨酸、色氨酸、缬氨酸。

表1 舞草种子油质体氨基酸种类及含量

单位:%

氨基酸种类	占种子油质体的百分比	占氨基酸总量的百分比
ASP 天门冬氨酸	0.90	6.94
THR 苏氨酸	0.30	2.29
SER 丝氨酸	0.31	2.36
GLU 谷氨酸	1.54	11.74
PRO 脯氨酸	1.64	12.50
GLY 甘氨酸	0.72	5.49
ALA 丙氨酸	0.24	1.83
CYS 胱氨酸	1.06	8.08
VAL 缬氨酸	0.10	0.76
MET 蛋氨酸	0.02	0.15
ILE 异亮氨酸	0.74	5.62
LEU 亮氨酸	0.71	5.41
TYR 酪氨酸	0.32	2.44
PHE 苯丙氨酸	0.48	3.64
LYS 赖氨酸	1.10	8.36
HIS 组氨酸	1.41	10.73
ARG 精氨酸	1.32	10.06
TRP 色氨酸	0.21	1.60
总量	13.12	100.00

无机元素是蚂蚁外骨骼的主要成分,同时它们对维持各器官的正常生理功能具有重要作用。蚂蚁必需的无机元素有钙、磷、铁、钾、铜、锰、镁、锌、碘等。从表2看出,舞草种子油质体中除碘外含有蚂蚁必需的其他8种无机元素。

表2 舞草种子油质体无机元素种类及含量

元素种类	S	P	K	Ca	Mg	Cu
占种子油质体的百分比/%	0.101 00	0.519 00	0.462 00	0.217 00	0.104 00	0.001 37
元素种类	Zn	Fe	Mn	B	Mo	
占种子油质体的百分比/%	0.001 48	0.039 10	0.007 90	0.000 97	0.000 10	

3 讨论

圆叶铺道蚁和布立毛蚁为舞草种子的主要传播者。圆叶铺道蚁数量较多,占样地搬运舞草种子蚂蚁数量的48%,且分布均匀;布立毛蚁尽管数量较少,但搬运速度快,搬运效率高。另据调查,圆叶铺道蚁还搬运热带雨林的关键物种——聚果榕的种子。

通常一种植物的种子都不仅由一种蚂蚁搬运,因此在做不同蚂蚁搬运效率比较时,仅仅在上、下午各观察30 min或1 h,以此来比较分析月间或季节间搬运效率的差异,就显得观察时间过短,且无法排除蚂蚁季节性活动差异所造成的影响。此外,国外还采用另一种试验方法,即将植物种子放置地上,采取措施只让蚂蚁进入,48 h或更长时间后统计被移走的种子数,得出月间或季节间搬运效率的差异。此方法虽较省时、省力,但无法统计不同蚂蚁种的搬运情况。本试验采用连续时间段观察,尽管比较费时、费力,但可以掌握搬运者一天各自的搬运率及其与日温变化的关系。通过各个季节的观察则可以掌握不同蚂蚁种群的活动规律并可比较其季节间种子搬运率的差异。

舞草种子成熟后,果荚自然裂开,种子仅能落到近植株的地上。成熟种子上附生的油质体能吸引蚂蚁搬运取食,而种子外壳较硬,蚂蚁等小昆虫无法取食,从而使舞草种子得以传播,种群得以繁衍。舞草种子成熟季节在每年的10—12月份,风和雨量都较少,因此,蚂蚁是其种子的主要传播者。

种子油质体是植物长期适应环境而形成的。植物在生长过程中,要耗费一定的能量和营养才能生成这些能吸引蚂蚁的物质,使其种子得以传播,从而繁衍其种群。在澳大利亚的某些干燥地区,约有24科,87属,约1500种植物的种子演化生成了这种油质体。油质体所占种子质量的比例变化较大,Stanislav曾对 *Asarum europaeum*,*Viola hirta*,*V. mirabilis*,*V. matutina* 及 *Chelidonium majus* 几种植物的种子油质体与种子的比例做过分析。舞草种子平均重6.985 mg/粒,油质体重0.248 mg/粒,油质体占种子总质量的3.55%,与上述5种植物相比,油质体所占种子质量的比例最小,这表明舞草植物为了种子的传播所付出的代价最小。

种子油质体富含蚂蚁生长发育所必需的氨基酸和无机元素。由于样品有限,其他蚂蚁生长所需的营养成分,如脂肪、碳水化合物、维生素等有待进一步分析。

原文刊载于《生态学报》,2001年第11期

多物种共存系统中拟水狼蛛对三种稻虫的捕食作用

李剑泉,赵志模

摘要: 选择稻田蜘蛛优势种拟水狼蛛(*Pirata subpiraticus*)和水稻主要害虫褐飞虱(*Nilaparvata lugens*)、白背飞虱(*Sogatella furcifera*)、黑尾叶蝉(*Nephotettix cincticeps*)组成多物种共存系统,运用二次通用回归旋转组合设计试验探讨蜘蛛对稻虫的控制作用,分析天敌与害虫间、害虫之间的相互作用关系,得到天敌对害虫的4个捕食量模型,其中拟水狼蛛对3种害虫的总捕食量模型为 $Y_4=28.143+6.375X_1+1.792X_2+1.625X_3-0.042X_4-4.754X_1^2-3.129X_2^2-1.629X_3^2-2.129X_4^2-0.188X_1X_2+0.063X_1X_3-0.438X_1X_4-0.188X_2X_3-0.438X_2X_4-1.438X_3X_4$。对该模型的主效分析表明,天敌自身密度对捕食量的影响最大;其次是害虫密度的影响,其中褐飞虱对捕食量的影响较大,白背飞虱次之,黑尾叶蝉最小。害虫间的互作分析表明,白背飞虱同褐飞虱间的交互效应对拟水狼蛛捕食作用的影响最强。

关键词: 共存系统;生物防治;种间关系;拟水狼蛛;稻虫

The Predatory Function of *Pirata subpiraticus* (Boes. et Str.) on Three Important Rice Insect Pests in a Multispecies Coexistent System

Li Jianquan, Zhao Zhimo

Abstract: By applying common rotational composite design of quadratic: regression, the predatory function of *Pirata subpiraticus* (Boes. et Str.), a dominant species of spiders in rice field in the suburb of Chongqing City, on three important rice insect pests, namely, the brown planthopper (BPH) *Nilaparvata lugens* Stal, white backed planthopper (WBPH) *Sogatella furcifera* (Horváth) and green rice leafhopper (GRH) *Nephotettix cincticeps* (Uhler) in a multi-species coexistent system was investigated. Four predation models derived from the analysis of interactions among predator and preys and within preys were developed, in which the total predation model of *Pirata subpiraticus* on BPH, WBPH and GRH was expressed as follows:

$Y_4 = 28.143 + 6.375X_1 + 1.792X_2 + 1.625X_3 - 0.042X_4 - 4.754X_1^2 - 3.129X_2^2 - 1.629X_3^2 - 2.129X_4^2 - 0.188X_1X_2 + 0.063X_1X_3 - 0.438X_1X_4 - 0.188X_2X_3 - 0.438X_2X_4 - 1.438X_3X_4$

Principal effect analysis of this model indicated that among the effects on predation of this spider, the density of predator itself was most important, and the density of insect pests was secondary, in which the effect order was BPH>WBPH>GRH. Analysis of the inter-species effects among the three insect pests on the predation of this spider showed that the action of WBPH×BPH was most powerful.

Key words: *Pirata subpiraticus*; predation; rice insect pest; interaction

褐飞虱(*Nilaparvata lugens*)、白背飞虱(*Sogatella furcifera*)和黑尾叶蝉(*Nephotettix cincticeps*)是重庆地区水稻的主要害虫,造成水稻产量损失一般在10%以上,危害重的达50%。近年来人们积极开展稻虫生物防治应用基础研究,探讨稻田蜘蛛的变动规律及其对害虫的自然控制机理,明确天敌捕食量的大小,建立捕食者-猎物系统的数学模型以预测天敌对稻虫的控制作用。稻田生态系统中,害虫之间、天敌之间以及天敌与害虫之间形成了极其复杂的网络关系。为揭示该系统中主要物种之间的相互关系,作者运用二次通用回归旋转组合设计试验,探讨了重庆地区稻田生态系统中拟水狼蛛对3种稻虫的控制效果。

1 材料与方法

1.1 试验设计

选择拟水狼蛛(X_1)、褐飞虱(X_2)、白背飞虱(X_3)和黑尾叶蝉(X_4)组成共存系统。根据1998—1999年对重庆市北碚区稻田蛛、虫的田间系统调查,确定试验因子的零水平和变化区间,按二次通用回归旋转组合设计,得试验方案(表1)。各参试因子编码值(X)与密度(d)的转换关系为:$X_1 = d_1/2 - 2$, $X_2 = d_2/15 - 2$, $X_3 = d_3/10 - 2$, $X_4 = d_4/5 - 2$。

表1 四物种共存系统试验设计结构矩阵与结果

编号No.	各因子编码值				各因子水平				观察值				拟合值			
	X_1	X_2	X_3	X_4	X_1	X_2	X_3	X_4	y_1	y_2	y_3	y_4	Y_1	Y_2	Y_3	Y_4
1	−1	−1	−1	−1	2	15	10	5	3	2	0	5	3.4	1.2	0.0	4.1
2	1	−1	−1	−1	6	15	10	5	10	5	0	15	11.5	6.1	0.4	18.0
3	−1	1	−1	−1	2	45	10	5	4	2	0	6	7.5	1.6	0.3	9.3
4	1	1	−1	−1	6	45	10	5	11	4	2	17	16.8	4.3	1.4	22.5
5	−1	−1	1	−1	2	15	30	5	3	4	0	7	4.5	5.5	0.6	10.5
6	1	−1	1	−1	6	15	30	5	9	11	1	21	9.8	13.7	1.2	24.6
7	−1	1	1	−1	2	45	30	5	6	5	1	12	8.3	6.2	0.5	15.0
8	1	1	1	−1	6	45	30	5	12	11	2	25	14.8	12.1	1.4	28.3
9	−1	−1	−1	1	2	15	10	15	4	2	0	6	5.0	3.0	0.8	8.7
10	1	−1	−1	1	6	15	10	15	10	4	2	16	12.3	6.2	2.4	20.8
11	−1	1	−1	1	2	45	10	15	5	2	1	8	8.8	2.7	0.7	12.1
12	1	1	−1	1	6	45	10	15	15	3	3	21	17.3	3.6	2.6	23.5
13	−1	−1	1	1	2	15	30	15	4	3	0	7	2.8	6.0	0.5	9.3
14	1	−1	1	1	6	15	30	15	7	10	2	19	7.3	12.5	1.9	21.7
15	−1	1	1	1	2	45	30	15	4	5	0	9	6.3	6.0	0.0	12.0
16	1	1	1	1	6	45	30	15	8	6	1	15	12.1	10.2	1.4	23.6
17	2	0	0	0	8	30	20	10	17	13	2	32	11.3	8.4	2.2	21.9
18	−2	0	0	0	0	30	20	10	0	0	0	0	0.0	0.0	0.0	0.0
19	0	2	0	0	4	60	20	10	19	10	2	29	9.8	8.4	1.0	19.2
20	0	−2	0	0	4	0	20	10	0	14	2	16	1.0	10.2	0.9	12.0
21	0	0	2	0	4	30	40	10	15	17	1	33	12.8	11.2	0.9	24.9
22	0	0	−2	0	4	30	0	10	23	0	1	24	17.0	0.4	1.0	18.4
23	0	0	0	2	4	30	20	20	15	11	2	28	11.8	6.2	1.5	19.5
24	0	0	0	−2	4	30	20	0	18	7	0	25	13.0	6.4	0.4	19.7

续表

编号No.	各因子编码值				各因子水平				观察值				拟合值			
	X_1	X_2	X_3	X_4	X_1	X_2	X_3	X_4	y_1	y_2	y_3	y_4	Y_1	Y_2	Y_3	Y_4
25	0	0	0	0	4	30	20	10	19	11	1	31	18.0	9.7	0.4	28.1
26	0	0	0	0	4	30	20	10	17	11	0	28	18.0	9.7	0.4	28.1
27	0	0	0	0	4	30	20	10	20	9	0	29	18.0	9.7	0.4	28.1
28	0	0	0	0	4	30	20	10	19	7	1	27	18.0	9.7	0.4	28.1
29	0	0	0	0	4	30	20	10	18	12	0	30	18.0	9.7	0.4	28.1
30	0	0	0	0	4	30	20	10	16	10	1	27	18.0	9.7	0.4	28.1
31	0	0	0	0	4	30	20	10	17	8	0	25	18.0	9.7	0.4	28.1

1.2 试验实施

选用口径35 cm、高50 cm的瓷钵，3月上旬填土灌水，4月26日盆栽水膜秧苗，每钵4丛，共栽31钵。常规管理，不施农药。在5月上旬准备好边长50 cm、高175 cm的木架，四周用细窗纱封闭，上封尼龙纱罩于瓷钵，并开口便于投放虫、蛛。试验于2000年6月24日—30日分2批进行，每批安排15或16个处理。清除基部死叶及水稻上其他昆虫与蜘蛛后，将从稻田中采回的蜘蛛，单头置于放有浸湿海绵的小三角瓶中，先饱食再饥饿1 d后，连同刚从稻田采回的害虫按试验设计放入盆栽水稻的笼罩内。2 d后观察害虫和天敌的剩余数、自然死亡数，计算害虫被捕食量，记入表1中。用y_1—y_3分别代表拟水狼蛛对褐飞虱、白背飞虱和黑尾叶蝉的捕食量，y_4代表拟水狼蛛对3种害虫的总捕食量。

1.3 数据处理

所有数据计算与处理均用DPS和Excel系统软件完成。

2 结果与分析

2.1 模型建立

将试验结果输入DPS平台,计算后得到共存系统的捕食量数学模型。用 Y_1—Y_4 分别代表褐飞虱、白背飞虱、黑尾叶蝉的被捕食量及三者总的被捕食量,X_1—X_4 分别代表拟水狼蛛和3种害虫的编码值。捕食量模型的各项系数(a为常数)如表2。其中拟水狼蛛对3种害虫总的捕食量模型为:

$Y_4 = 28.143 + 6.375X_1 + 1.792X_2 + 1.625X_3 - 0.042X_4 - 4.754X_1^2 - 3.129X_2^2 - 1.629X_3^2 - 2.129X_4^2 - 0.188X_1X_2 + 0.063X_1X_3 - 0.438X_1X_4 - 0.188X_2X_3 - 0.438X_2X_4 - 1.438X_3X_4$。

表2 四物种共存系统中各捕食量模型参数的系数矩阵

参数	模型				参数	模型			
	Y_1	Y_2	Y_3	Y_4		Y_1	Y_2	Y_3	Y_4
a	18.000*	9.714*	0.429*	28.143*	X_4^2	-1.406*	-0.856*	0.132	-2.129*
X_1	3.458*	2.292*	0.625*	6.375*	X_1X_2	0.313	-0.563	0.063	-0.188
X_2	2.208*	-0.458	0.042	1.792*	X_1X_3	-0.688*	0.813*	-0.063	0.063
X_3	-1.042*	2.708*	-0.042	1.625*	X_1X_4	-0.188	-0.438	0.188	-0.438
X_4	-0.292	-0.042	0.292*	-0.042	X_2X_3	-0.063	0.063	-0.188	-0.188
X_1^2	-3.406*	-1.481*	0.132	-4.754*	X_2X_4	-0.063	-0.188	-0.188	-0.438
X_2^2	-3.156*	-0.106	0.132	-3.129*	X_3X_4	-0.813*	-0.313	-0.313*	-1.438*
X_3^2	0.781*	-0.981*	0.132	-1.629					

注:*表示 $a=0.10$ 的显著水平。

试验表明,拟水狼蛛对供试的3种害虫在嗜食性上存在差异。根据捕食量模型 Y_1—Y_3(表2),固定蛛、虫的编码值均为零水平($X_1=4, X_2=30, X_3=20, X_4=10$),计算拟水狼蛛对褐飞虱、白背飞虱和黑尾叶蝉的捕食量分别为4.500、2.429和0.107头,由此看出拟水狼蛛对褐飞虱的捕食量最大,对白背飞虱的捕食量次之,对黑尾叶蝉的捕食量最小。

2.2 主效分析

二次通用回归旋转组合设计满足试验的正交性,模型中各项效应线性可加,偏回归系数间彼此独立。因此固定其他变量于零水平,从模型 Y_4 可得到拟水狼蛛对 3 种害虫捕食量的主效作用方程;通过对主效方程求导,可得到密度作用方程和作用系数(表3)。

由表3可知,随着天敌和害虫密度的增加,捕食量呈抛物线增减。天敌蜘蛛和3种害虫都是在低密度时,随着密度的增加捕食量也随之增加,其中增加最快的是拟水狼蛛,褐飞虱次之,白背飞虱较慢,黑尾叶蝉最慢;当增加到某一密度时,捕食量达到最大值,超过该密度值捕食量反而下降,其中褐飞虱下降最快,黑尾叶蝉和拟水狼蛛次之,白背飞虱下降最慢。另外,4种共存因子的作用系数均为负值,其中 X_1 的系数绝对值最大,说明拟水狼蛛的种内干扰对捕食量的影响最为突出;而3种害虫的作用系数绝对值较小,说明其个体间的干扰对捕食量的影响相对较弱。

表3 模型 Y_4 的单因子主效分析(其他因子为零水平)

参数	主效作用方程	密度作用方程	作用系数
X_1	$28.143+6.375X_1-4.754X_1^2$	$6.375-9.509X_1$	-9.509
X_2	$28.143+1.792X_2-3.129X_2^2$	$1.792-6.259X_2$	-6.259
X_3	$28.143+1.625X_3-1.629X_3^2$	$1.625-3.259X_3$	-3.259
X_4	$28.143-0.042X_4-2.129X_4^2$	$-0.042-4.259X_4$	-4.259

2.3 互作分析

把模型 Y_4 的其他变量固定在零水平上,可得到6个两物种的互作方程。作出它们交互作用的反应面(图1),可以分析天敌与害虫之间、一种害虫与另一种害虫之间的相互干扰对捕食量的影响。

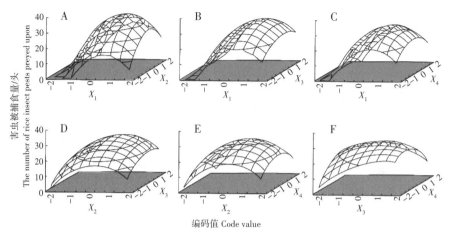

图 1 多物种共存系统中各物种之间交互作用下害虫被捕食量反应面

注:X_1 拟水狼蛛 Pirata subpiraticus, X_2 褐飞虱 Nilaparvata lugens, X_3 白背飞虱 Sogatella furcifera, X_4 黑尾叶蝉 Nephotettix cincticeps。

图 1 直观地反映了各因子间的互作效应对捕食量的影响。A—C 表明,拟水狼蛛(X_1)与 3 种害虫的交互效应对捕食量的影响趋势基本一致。它们都是天敌在较低密度下捕食量很小,随着自身密度和害虫密度的增加捕食量迅速增加,当密度增加到较高时,捕食量又随密度的增加而减少,但维持相当水平,因而蜘蛛仍能发挥其应有的控制作用。比较 A、B、C 可知,A 相对较拱,表明 X_1 与 X_2(褐飞虱)的互作对捕食量增减的影响较大,X_1 与 X_4(黑尾叶蝉)的互作影响次之,X_1 与 X_3(白背飞虱)的互作影响相对较小。D—F 表明,害虫被捕食量均是随自身密度的增加呈抛物线性增减,但 E、D 曲面相对较拱,表明 X_2 与 X_3、X_2 与 X_4 的互作对捕食量增减的影响较大,而 F 曲面相对较平,表明 X_3 与 X_4 的互作对捕食量增减的影响较小。

3 结论与讨论

四物种共存系统中,供试天敌拟水狼蛛最喜欢捕食褐飞虱,其次是白背飞虱,对黑尾叶蝉的捕食量很小。主效分析表明,蜘蛛和害虫各自的密度对天敌的捕食量有明显影响,特别是拟水狼蛛。影响的一般趋势都是在低密度时,捕食量随密度的升高而增加,在较高密度时随密度的上升而降低,但仍维持在相当水平。这表明蛛、虫种内个体间存在对捕食的干扰关系,特别是拟水狼蛛种内干扰对捕食量的影响最为突出,而3种害虫个体间的干扰对捕食量的影响相对较弱。

互作分析表明,随着天敌和害虫密度的增加,害虫被捕食量呈抛物线增减。从互作反应面(图1)可知,褐飞虱同白背飞虱间的相互干扰对捕食量的影响最强,褐飞虱同黑尾叶蝉的影响次之,白背飞虱同黑尾叶蝉的影响较弱。比较这些害虫可以得出,个体大、活动能力强、生态位宽,其相互干扰越强;反之其干扰作用较弱。Iwasa等分析了密度相关、逆密度相关、密度无关捕食率的形成原因。按本文模型 Y_4 计算,当天敌密度(1水平)为6头,害虫密度(均在零水平)为60头时,天敌对害虫的捕食率最高,如果害虫密度超过这个值,则与捕食率呈逆密度制约关系,因此现有天敌难以有效控制害虫而必须采用其他的防治措施。

在多物种共存系统中,天敌间、害虫间以及天敌与害虫间存在着非常复杂的种间及种内关系,这种关系不仅影响天敌和害虫的生长发育和繁殖,而且也制约着天敌对害虫的捕食效能。因此进一步探讨农田生态系统中多个物种之间的相互关系,对于害虫的持续控制具有十分重要的意义。

原文刊载于《植物保护学报》,2002年第1期

瓜类蔬菜昆虫群落(包括蛛形纲和软体动物)的研究Ⅲ.群落的数量动态

陈宏，靳阳，赵志模

摘要：本文报道了瓜类昆虫群落(包括蛛形纲和软体动物门中的一些有害生物及天敌)的数量动态，从两个方面进行了分析描述。①采用最优分割分析各瓜类昆虫群落的发展阶段；②采用主分量分析方法探讨各昆虫种群对群落特征的作用。

关键词：瓜类蔬菜；昆虫群落；数量动态

Studies on Insect Communities (Including Arachnida and Mollusca) of Gourd Vegetables Ⅲ. Quantitative Dynamics of the Communities

Chen Hong, Jin Yang, Zhao Zhimo

Abstract: This paper reports the quantitative dynamics of insect communities (including some pests and natural enemies belonging to Arachnida and Mollusca). Decription and analysis in two aspects: (1) The seasonal pattern of insect community in gourd vegetables was divided objectively into some related and different stages by the method of optimal sorting. (2) The principal component analysis method was used to examine the effects of every insect species on characteristics of community.

Key words: gourd vegetables; insect communities; quantitative dynamics

0 引言

在菜地昆虫群落研究方面，王金福等调查研究了杭州郊区主要蔬菜害虫群落结构，赵志模、刘映红等1994年报道了重庆市郊不同种植制度菜地昆虫群落的结构等内容。本项研究的目的就在于从昆虫群落的水平弄清重庆市郊苦瓜、黄瓜、丝瓜、南瓜和冬瓜五种主要瓜类蔬菜昆虫群落的发展规律。主要调查重庆市郊,为制定和实施不同瓜类害虫综合防治提供依据。

1 调查方法

本文主要对逐次资料反映出来的群落结构在时间上的变化进行详细分析描述。各瓜类昆虫群落阶段分别采用最优分割法分析,各昆虫在群落中的作用采取主分量分析法分析,涉及的各种具体方法,在结果与分析中一并阐述。

2 结果与讨论

2.1 瓜类主要昆虫群落的发展阶段

分别将各种瓜类上的主要害虫和天敌作为一个整体,将物种及其数量作为时间样本的属性,采用最优分割分析各瓜类昆虫群落的发展阶段。

将五种瓜上昆虫系统调查的资料分别整理得出如下数据矩阵(表略)

$$X_{p \times n} = \begin{bmatrix} X_{11} & X_{12} & \cdots & X_{1n} \\ X_{21} & X_{22} & \cdots & X_{2n} \\ \vdots & \vdots & & \vdots \\ X_{p1} & X_{p2} & \cdots & X_{pn} \end{bmatrix}$$

式中$X_{ij}(i=1,2,\cdots,p;j=1,2,\cdots,n)$表示第$j$个样本(月份)第$i$个属性(物种)的观察值(物种数量)。

如果对于n个样本进行分割而不破坏其样本的固有顺序,则可将其样本分为二段、三段……直至$n-1$段,并分别二分割、三分割……K的分割($K=n-1$)。方式有C_{n-1}^2种,现要求在各种分割中找出一种分割方式,让其段内的差异最小,而段间的差异最大。该种分割则称为最优分割。显然对于2,3,…,k分割都有其自身的最优分割方式,分别称为最优二分割、最优三分割……最优K分割。

本文采用离差平方和 d_{ij} 表示段的差异 $d_{ij}=\sum_{a=1}^{p}\sum_{k=i}^{j}[X_{ak}-\bar{X}_{a(i,j)}]^2$

式中 $\bar{X}_{a(i,j)}=\dfrac{1}{j-i+1}\cdot\sum_{k=i}^{j}X_{ak}$

它是第 a 个属性，第 i 个样方变差的平均值，d_{ij} 越小，意味着子段内各月昆虫组成的数量越相近。反之差异越大。可以证明，各子段差异之和越小，则段间的差异大。因此在各分割中即可选择段内变异之和最小的方式作为最优分割，结果见表1。

表1 瓜类昆虫群落按月分段的最优分割

瓜类	分割数X	最优分割结果/月份							$S_{段内}$
苦瓜	2	4	5	6	7	8	9	10	8.890
	3	4	5|	6	7	8	9	10	4.080
	4	4	5|	6	7|	8	9	10	2.670
	5	4|	5|	6	7|	8	9	10	1.310
	6	4|	5|	6	7|	8	9|	10	0.630
黄瓜	2	4	5	6	7|	8	9	10	12.100
	3	4	5	6	7|	8	9	10	5.930
	4	4	5	6|	7|	8	9	10	1.520
	5	4|	5	6|	7|	8	9	10	0.012
	6	4|	5	6|	7|	8	9|	10	0.000
丝瓜	2	4	5|	6	7	8	9	10	13.120
	3	4	5	6|	7	8	9|	10	8.270
	4	4	5	6|	7|	8	9|	10	4.870
	5	4	5	6|	7|	8	9|	10	1.970
	6	4	5	6|	7|	8	9|	10	0.660
冬瓜	2	4	5	6	7|	8	9	10	14.330
	3	4	5|	6	7|	8	9	10	7.520
	4	4	5|	6|	7|	8	9	10	3.790

续表

瓜类	分割数 X	最优分割结果/月份						$S_{段内}$
冬瓜	5	4	5|	6|	7|	8	9　10	1.990
	6	4|	5|	6|	7|	8|	9　10	0.360
南瓜	2	4	5	6|	7	8	9　10	13.470
	3	4	5|	6|	7	8	9　10	8.340
	4	4	5|	6|	7	8|	9　10	4.760
	5	4|	5|	6|	7	8|	9　10	2.750
	6	4|	5|	6|	7|	8|	9　10	0.870

在最优分割中选择几分割来划分阶段尚无一定之规，总的来说，主要应视分段的生物学背景和样本数的多少而定。郭依泉曾对柑桔群落周年12个月的样本进行最优分割，并根据各分段内变差曲线的拐点位于最优四分割处而将其划分为4个阶段，鉴于本文所研究的瓜类生育期较短，调查月份4—10月，仅有7个样本，所以对5种瓜类群落均采取最优三分割来划分阶段。

由表1看出：苦瓜、黄瓜和冬瓜上的三个昆虫群落划分为4—5月，6—7月和8—10月三个阶段；丝瓜划分为4—6月，7—9月和10月三个阶段；南瓜划分为4—5月，6月和7—10月三个阶段。

一般来说，按时间序列划分阶段时其第一阶段往往反映了群落发展的初期，4—5月在苦瓜、黄瓜、冬瓜以及南瓜上，4—6月在丝瓜上昆虫的种类及数量较少。正是各昆虫种开始形成种群的时期，从瓜的生育期看正值苗期或伸蔓期，从气候看反映出春末夏初的特点。这个阶段经历的时间在丝瓜上相对较长。第二阶段正是群落发展的盛期，昆虫种类最多、数量最大，这个阶段南瓜昆虫群落历时最短，仅在6月，丝瓜昆虫群落历时最长，且时间较迟（在7—9月）。第三阶段反映了群落的消退，这主要受瓜类生育期的影响，丝瓜的第三阶段仅在10月，群落消退较迟，而南瓜昆虫群落消退较早，且持续期长（在7—10月），而苦瓜、黄瓜和冬瓜昆虫群落的消退较丝瓜为早，较南瓜为迟（在8月）。

2.2 瓜类昆虫群落中各虫种对群落特征的作用

各虫种对群落特征的作用采用主分量分析方法进行。因为瓜类植物均属于葫芦科植物,几种瓜上的昆虫群落在物种组成及其数量特征上亦具有许多共同之处,不同瓜类的种植地块或镶嵌或隔离尚远,但因瓜类昆虫食性和嗜好和一定的迁徙能力,可以把几种瓜类作物上的昆虫群落作为一个整体来看,探讨整个瓜类昆虫群落在时间序列中各种昆虫对群落特征的作用。其全部过程在计算机 Statgraphics 软件上完成。

表2列出了瓜类昆虫群落7个月份(样本),21个物种(属性)的数据。各主分量的特征根及贡献率如表3。

由表3看出,将原来21维属性(物种)的数据降到二维的主分量排序时,其保留的信息量达79.861 2%,损失信息量仅20.138 8%,因此笔者仅以两个主分量分析各种昆虫在群落中的作用。

表2 瓜类昆虫群落4—10月各种昆虫的数量

单位:个

物种	4月	5月	6月	7月	8月	9月	10月
棉蚜 *Aphis gossypii* Glover	638	3 195	8 788	1 381	34	796	5 367
朱砂叶螨 *Tetranychus cinnabarinus* Boisduval	215	7 378	17 419	5 672	475	1 021	67
温室白粉虱 *Trialeurodes vaporariorum*(Westwood)	0	117	411	663	452	164	14
黄守瓜 *Aulacophora femoralis chinensis* Weise	25	164	131	385	402	362	5
黄足黑守瓜 *Aulacophora lewisii* Baly	0	48	26	158	165	544	143
瓜亮蓟马 *Thrips flevas* Schrank	7	272	1 117	1 917	335	91	0
美洲斑潜蝇 *Liriomyza sativae* Blanchard	5	19	21	106	247	397	13

续表

物种	4月	5月	6月	7月	8月	9月	10月
双纹斑叶蝉 *Erythroneura limbata*(Matsumura)	0	12	37	118	72	19	0
假穴环肋螺 *Plectotropis pseudopatula* Moellendorff	0	73	158	434	100	28	4
灰巴蜗牛 *Bradybaena ravida*(Benson)	0	0	86	76	28	11	3
大造桥虫 *Ascotis selenaria* Schiffermuller et Denis	0	19	175	21	0	30	34
瓜绢野螟 *Diaphania indica*(Saunders)	0	13	11	24	3	30	57
野蛞蝓 *Agriolimax agrestis*(Linnaeus)	1	21	144	16	1	0	0
锯齿射带蜗牛 *Laeocathaica prionotropis*(Moellendorff)	0	21	18	0	0	0	0
细角瓜蝽 *Megymenum gracilicorne* Dallas	0	27	50	16	1	0	0
黄曲条跳甲 *Phyllotreta striolata*(Fabricius)	0	42	137	97	47	47	10
小花蝽 *Orius minutus*(linnaeus)	0	0	147	394	141	20	0
草间小黑蛛 *Erigonidium graminicolum*(Sundevall)	0	61	162	157	58	43	25
菜少脉蚜茧蜂 *Diaeretiella rapae* Curtis	0	42	145	42	0	0	25
异色瓢虫 *Harmonia axyridis*(Pallas)	0	4	29	66	1	6	17
八斑球腹蛛 *Theridion octomaculatum* Boes. et Str.	0	0	11	63	17	9	1

表3　各主分量的特征根及贡献率

序号	特征根	贡献率/%	累积贡献率/%
1	10.618 800	50.565 82	50.565 82
2	6.152 020	29.295 33	79.861 15
3	1.906 340	9.077 83	88.938 98
4	1.588 330	7.563 50	96.502 48
5	0.533 925	2.542 50	99.044 98
6	0.200 554	0.955 02	100.000 00

主分量是原来所有属性的线性组合,它虽然不能解释成单个属性的作用,然而仍然有可能通过考察主分量对应的特征向量来说明原来各属性对主分量作用的大小。特征向量中各元素在-1到1之间,它们的平方和等于1。因此当特征向量中某元素的平方值越大时,则这个属性对主分量的作用亦越大。另外我们亦可用属性对主分量的负荷量来反映属性对主分量的作用。属性i对第j个主分量的负荷量:

$$\lambda_{ji} = \bar{\lambda}_j \cdot U_{ji} (i,j=1,2,\cdots,p)$$

式中λ_j是第j个主分量的特征根,U_{ji}是第j个主分量对应的特征向量中第i个属性元素。显然负荷量综合考虑了特征根和特征向量。前两个主分量对应的特征向量及各属性负荷量见表4。

由表4可以分析在时间序列上引起瓜类昆虫群落特征变化的主要种类。对第一主分量(占信息量的50.6%),贡献较大的是草间小黑蛛、灰巴蜗牛、黄曲条跳甲、瓜亮蓟马、温室白粉虱、假穴环肋螺、朱砂叶螨、棉蚜,其负荷量在0.801 9~0.992 9之间,而贡献较小的是黄守瓜、黄足黑守瓜、美洲斑潜蝇和瓜绢野螟,其负荷量仅为0.154 5~0.297 3,结合表2分析上述贡献率较大和贡献较小的昆虫发现,贡献较大的昆虫,其发生高峰期基本都在6—7月,而贡献较小的昆虫的发生高峰期都在7月以后,可以认为第一主分量基本代表了发生期较早的昆虫种类。对第二主分量(占信息量的29.3%)贡献较大的昆虫有黄守瓜、棉蚜、锯齿射带蜗牛和八斑球腹蛛,其负荷量在0.714 3~0.779 1之间。其中棉蚜、锯齿射带蜗牛负荷量为负值,说明其对群落特征的影响与负荷量为正的昆虫有完全相反的趋势。

主分量分析实质上是一种排序技术,本文是将时间序列的月份作为实体,在以物种数量为属性的坐标轴的空间中,按其相似关系排列出来,从而阐明瓜类昆虫群落在时间序列上的特征。表5给出了4—10月在第一、二分量上的排序坐标(见图1、图2)。

表4　第一、二主分量的特征向量及各属性负荷量

物种	第一主分量		第二主分量	
	特征向量	负荷量	特征向量	负荷量
棉蚜 *Aphis gossypii* Glover	0.246 112 0	0.801 9	−0.312 931 0	−0.776 2
朱砂叶螨 *Tetranychus cinnabarinus* Boisduval	0.247 857 0	0.807 7	−0.221 681 0	−0.549 8
温室白粉虱 *Trialeurodes vaporariorum*(Westwood)	0.251 346 0	0.819 0	0.207 481 0	0.514 6
黄守瓜 *Aulacophora femoralis chinensis* Weise	0.091 249 1	0.297 3	0.314 117 0	0.779 1
黄足黑守瓜 *Aulacophora lewisii* Baly	−0.073 994 6	−0.241 1	0.202 790 0	0.502 8
瓜亮蓟马 *Thrips flevas* Schrank	0.282 878 0	0.921 8	0.127 735 0	0.316 8
美洲斑潜蝇 *Liriomyza sativae* Blanchard	−0.056 117 7	−0.182 9	0.242 732 0	0.602 1
双纹斑叶蝉 *Erythroneura limbata*(Matsumura)	0.208 092 0	0.678 1	0.281 819 0	0.699 0
假穴环肋螺 *Plectotropis pseudopatula* Moellendorff	0.250 497 0	0.816 3	0.202 203 0	0.501 5
灰巴蜗牛 *Bradybaena ravida*(Benson)	0.296 236 0	0.965 3	0.035 843 5	0.088 9
大造桥虫 *Ascotis selenaria* Schiffermuller et Denis	0.203 371 0	0.662 7	−0.248 203 0	−0.615 6

续表

物种	第一主分量		第二主分量	
	特征向量	负荷量	特征向量	负荷量
瓜绢野螟 *Diaphania indica*(Saunders)	−0.047 414 0	−0.154 5	0.015 63 0	0.038 8
野蛞蝓 *Agriolimax agrestis*(Linnaeus)	0.226 046 0	0.736 7	−0.246 155 0	−0.610 5
锯齿射带蜗牛 *Laeocathaica prionotropis*（Moellendorff）	0.120 291 0	0.391 9	−0.288 008 0	−0.714 3
细角瓜蝽 *Megymenum gracilicorne* Dallas	0.239 648 0	0.780 9	−0.230 111 0	−0.570 8
黄曲条跳甲 *Phyllotreta striolata*(Fabricius)	0.293 125 0	0.955 2	−0.025 655 4	−0.066 3
小花蝽 *Orius minutus*(linnaeus)	0.243 079 0	0.792 1	0.231 518 0	0.574 2
草间小黑蛛 *Erigonidium graminicolum*(Sundevall)	0.304 682 0	0.992 9	0.021 549 4	0.053 4
菜少脉蚜茧蜂 *Diaeretiella rapae* Curtis	0.243 861 0	0.794 7	−0.239 727 0	−0.594 6
异色瓢虫 *Harmonia axyridis*(Pallas)	0.242 045 0	0.788 7	0.140 820 0	0.349 3
八斑球腹蛛 *Theridion octomaculatum* Boes. et Str.	0.201 884 0	0.657 9	0.287 979 0	0.714 3

由图1看出,以瓜类昆虫群落中物种数量表示的月份为其实体时,在第一主分量上表现为一条上拱的曲线,曲线从4月开始上升,6—7月达到高峰,其后逐渐下降。这一状况显然可以由对第一主分量贡献最大的几种昆虫的发生情况予以解释。月份实体在第二主分量上呈现波动的曲线,而其高峰在7—8月,从对第二主分量贡献最大的物种种类看,其负荷量有正有负,大小无序。因此曲线的波动是不可避免的。

表5　不同月份在第一、二主分量上的坐标值

主分量	4	5	6	7	8	9	10
第一主分量	-3.092 2	-0.739 2	4.930 5	4.259 9	-0.779 9	-2.122 0	-2.457 2
第二主分量	-0.926 6	-1.867 7	-3.463 9	3.626 8	2.075 7	1.478 2	-0.520 5

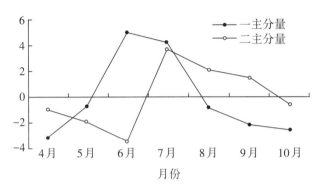

图1　不同月份在第一、二主分量上的一维排序

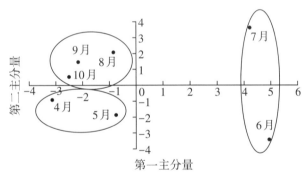

图2　不同月份在第一、二主分量上的二维排序

由图2看出，月份实体在以第一、二主分量为坐标的平面上沿着每主分量（横坐标）的变差较大，即从-3.092 2到4.930 5，而沿着第二主分量（纵坐标）的变差较小，为-3.463 9到3.626 8，这说明第一主分量信息含量（50.6%），明显高于第二主分量（信息含量29.3%）。从月份实体在二维平面上的信息量看，如果不破坏月份固有顺序性，6月和7月可划为一个组，它主要是对于第一主分量的。4、5月可成为一个组，8、9、10月可分为一个组，它们主要是对于第二主分量的。这种划分作为瓜类昆虫群落总体的发展阶段与前述利用最优分割探讨各瓜类昆虫群落发展的阶段相吻合。

采用最优分割方法和主分量分析两种方法进行分析,结果相吻合,昆虫群落的发展从时间上(除少数种类外)可分三个阶段,4—5月是发展的初级阶段,6—7月是发展的高峰期,8—10月是发展的末期,说明瓜类昆虫群落的发展有一定规律。

上述两种方法对瓜类昆虫群落在时间序列上的发展动态进行了系统分析,结果表明,瓜类昆虫群落数量随时间的变化有一定规律性。我们通过系统分析,弄清不同瓜类昆虫群落数量动态的发展规律、形成过程,对于了解害虫与天敌间的相互关系、天敌的保护利用及制定以昆虫群落为中心的害虫综合治理策略和方案具有重要参考价值。

原文刊载于《南开大学学报(自然科学版)》,2004年第1期

不同微生境中泽兰实蝇寄生对紫茎泽兰有性繁殖的影响

王文琪,王进军,赵志模

摘要:泽兰实蝇 *Procecidochares utilis* Stone 是紫茎泽兰重要的专性寄生性天敌,有性繁殖是紫茎泽兰长距离扩散蔓延的主要途径。为评价泽兰实蝇在不同微生境中对紫茎泽兰的控制作用,在四川省德昌县选择沟谷地、撂荒地、果园和马尾松林地4种微生境,采用田间调查结合盆栽试验的方法,分析了泽兰实蝇对紫茎泽兰的寄生规律以及有性繁殖的影响。结果表明,泽兰实蝇在不同微生境中对紫茎泽兰寄生率的高低依次为沟谷地31.7%、马尾松林地28.1%、果园26.1%和撂荒地13.5%,这一顺序反映了泽兰实蝇的寄生率有随微生境中土壤含水量升高而增加的趋势。泽兰实蝇寄生后,紫茎泽兰单位面积上的花序数和种子千粒重均显著下降,而单个花序内的种子数无显著变化。此外泽兰实蝇寄生后紫茎泽兰种子的长度和宽度、种子活力、萌发率和出苗率均呈下降趋势,降幅分别在8.1%~14.9%、14.7%~18.4%、29.8%~37.9%、28.6%~43.4%和58.7%~69.1%之间,说明泽兰实蝇寄生影响了种子的生命力,从而对紫茎泽兰的扩散蔓延起到一定的抑制作用。

关键词:泽兰实蝇;寄生;微生境;紫茎泽兰;有性繁殖

Effects of Parasitizing on the Sexual Reproduction of *Eupatorium adenophorum* Sprengel by *Procecidochares utilis* Stone at Different Microhabitats

Wang Wenqi, Wang Jinjun, Zhao Zhimo

Abstract: *Procecidochares utilis* Stone is an important specific natural enemy parasite on *Eupatorium adenophorum* Sprengel, and the sexual reproduction is the main approach for the invasion and expansion of *E. adenophorum*. In order to evaluate the control efficiency of *Procecidochares utilis* against *Eupatorium adenophorum* in different microhabitats, four different microhabitats, namely valley, wasteland, orchard, and woodland of pine trees, were selected to investigate the parasitic pattern and its influence on sexual reproduction of *E. adenophorum* in Xichang County,

Sichuan Province, using the method of field survey and pot experiments. The results showed that *Procecidochares utilis* had significant selectivity for microhabitats of parasitizing. The parasitizing rate in valley microhabitat (31.7%) was the highest, followed by woodland of pine tree (28.1%), orchard (26.1%) and wasteland (13.5%). The sequence of parasitizing rate implied that the tendency of parasitizing rate increases with the increase of water content of soils in the microhabitats. After parasitizing, the amount of anthotaxy and the 1 000 grain weight of *Eupatorium adenophorum* decreased significantly in different microhabitats, while the effects on the average seed amount within the anthotaxy were not significant. The seed length and width, the seed vigor, germination rate and emergence rate of *E. adenophorum* were all reduced significantly, and the decrease extent were 8.1%—14.9%, 14.7%—18.4%, 29.8%—37.9%, 28.6%—43.4% and 58.7%—69.1%, respectively. The results suggested that the parasite by *Procecidochares utilis* had lowered the degree of the seed vitality, thus have the inhibiting function to invasion and expansion of *Eupatorium adenophorum*.

Key words: *Procecidochares utilis*; parasitizing behavior; microhabitat; *Eupatorium adenophorum*; sexual reproduction

紫茎泽兰 *Eupatorium adenophorum* Sprengel 是一种多年生的有毒植物,以种子繁殖为主,亦具很强的无性繁殖力。由于它的生态适应性广、生长速度快、繁殖能力强,大约从20世纪40年代由中缅边境传入我国云南后,在半个多世纪里通过侵入、定居、适应到扩展,现已分布到我国西南广大地区,对农林牧业造成了巨大危害。紫茎泽兰成熟母株每年2—4月份可产生数量庞大(24万~28万粒/m^2)、活力指数高并具有冠毛的微小种子,这些种子不仅可借助西南季风向东向北扩散,而且还可随人、畜活动,依附交通工具或随河水漂移进行远距离传播蔓延,因此紫茎泽兰强大的有性繁殖能力是其快速蔓延尤其是长距离扩散的主要原因。

泽兰实蝇 *Procecidochares utilis* Stone 是紫茎泽兰重要的专性寄生性天敌。美国、澳大利亚、新西兰和印度等国家曾先后利用该虫进行生物防治来控制紫茎泽兰的危害。我国于1984年首次从西藏聂拉木县将泽兰实蝇引进云南部分地区用于控制紫茎泽兰,现已在该省5 000 km^2 的范围内逐步定殖,随后被引入四川、贵州等地并形成自然种群。研究发现,泽兰实蝇通过幼虫蛀入紫茎泽兰幼嫩茎枝端部,蛀食并刺激被害部位形成膨大虫瘿,严重抑制紫茎泽兰的萌芽率、分枝数、株高和种子产量,并抑制其光合作用及生物量的增加和分配,使节间缩短甚至枯死等。国内外在泽兰实蝇的利用方面虽进行了较多的研究,但多集中于泽兰实蝇寄生后对紫茎泽兰生长发育的影响及控制效果方面,而对有性繁殖尤其是定量研究的报道较少,对在自然条件下不同微生境中有性繁殖影响的研究更是缺乏。鉴于此,作者研究了不同微生境中泽兰实蝇寄生对紫茎泽兰有性繁殖的影响,以期探讨泽兰实蝇的寄生规律及对紫茎泽兰有性繁殖的影响,为紫茎泽兰的综合防除提供理论依据。

1 材料与方法

1.1 研究地点

研究地点设在四川省西南部攀西经济区的德昌县境内,东经102°16′,北纬21°56′,平均海拔1 430 m,年均气温18 ℃,常年日照时数不少于2 147 h,≥10 ℃积温5 120～5 426 ℃,年均降雨量1 049 mm,90%以上的降雨集中于6—10月,自然条件非常适合紫茎泽兰的生长。

1.2 研究方法

通过对土壤养分、水分、光照等生境条件的调查,在沿公路线25 km范围内选择具有代表性的4种微生境,分别为沟谷区(土壤湿度:26.38%±1.78%;全氮:0.083%±0.004%;磷:29.73 mg/kg±0.67 mg/kg)、撂荒地(土壤湿度:13.62%±0.78%;全氮:0.062%±0.002%;磷:26.13 mg/kg±0.49 mg/kg)、管理粗放的枇杷果园(土壤湿度:17.58%±1.18%;全氮:1.014%±0.002%;磷:41.30 mg/kg±2.11 mg/kg)和马尾松纯林地(土壤湿度:18.92%±1.18%;全氮:0.094%±0.002%;磷:37.30 mg/kg±1.19 mg/kg)。每一生境设置南—北走向、面积为8 m×30 m的样地各3块。各微生境样地的海拔在1 380～1 450 m之间,土壤类型均为山地红壤土。

由于多年生紫茎泽兰成株克隆生殖能力很强,主枝又可产生众多分枝,因此对其有性繁殖特性,如花序数、种子数等,按样方面积调查。在紫茎泽兰开花期的2—3月份,每个样地中随机选取1 m×1 m的样方15个,共计 180 个样方,调查泽兰实蝇的寄生率;随后再在每块样地中随机选取 50 cm×50 cm 的小样方 20 个,共计 240 个小样方,调查并计算单位面积上被寄生和未被寄生的花序总数;从每一个小样方中随机选取被寄生和未被寄生的头状花序各 15 个,共计 3600 个花序,计数每一个花序中的种子数量;从紫茎泽兰种子开始成熟的 3 月中旬起到 4 月上旬,在不同微生境中分别采集大量被寄生和未被寄生的花序所生产的种子,在室温下放置 1 月使其干燥,再测定其千粒重、长度和宽度。

紫茎泽兰种子活力测定采用四唑法;种子发芽率测定采用纸皿法,每皿种子 50 粒,10 次重复,在温度 25 ℃±1 ℃、光照 12 h 和黑暗 12 h 条件下观察,以胚根伸出种子长 1/2 定为发芽,试验期间每天统计发芽数,种子连续 5 d 无萌发则停止观察。

种子出苗率采用盆播试验测定,分别将采自不同微生境中的种子播在直径 18 cm、深 14 cm、盛有菜园红壤土的塑料盆钵中,每盆播 100 粒,行距 2 cm,株距 1 cm,深 0.5 cm,每处理 10 次重复,在一层遮阳网下进行萌发试验。试验期间每天观察记录种子的萌发和存活情况,直至苗高达 3 cm 左右为止。

所得数据用 SPSS 10.0 for Windows 进行分析。

2 结果与分析

2.1 不同微生境中泽兰实蝇寄生对紫茎泽兰花序及其种子数量的影响

(1)泽兰实蝇的寄生率:在 4 个微生境中,沟谷地中泽兰实蝇寄生率最高,达 31.7%,显著高于其余 3 种生境(表 1)。泽兰实蝇在果园和林地中的寄生率差异不显著,撂荒地中的寄生率最低,可能与撂荒地中的水肥条件较差有关。

(2)单位面积内紫茎泽兰的平均花序数:未寄生的紫茎泽兰花序数在沟谷区最多,撂荒地最少且与其他 3 种生境差异显著;泽兰实蝇寄生后各生境紫茎泽兰花序数均显著下降,其中在沟谷区生境中的降幅最大,达 25.8%,其后依次为林地、果园和撂荒地(表 1)。

(3)单个花序内的种子平均数量:单个花序内种子平均数量在未被寄生情况下以管理粗放的果园最高,撂荒地最低,且种子数量多少与生境中的土壤肥力呈正相关(表1)。泽兰实蝇寄生后各生境单个花序内种子数量均有不同程度的下降,其中以果园降幅最大,其后依次为林地、沟谷区和撂荒地,但各生境的降幅均不显著。

(4)单位面积中紫茎泽兰种子的平均数量:泽兰实蝇寄生对紫茎泽兰种子的产出量抑制作用明显,4种生境中寄生后种子数量均显著下降,并且这种抑制作用有随土壤湿度增加而增强的趋势(表1)。

表1 泽兰实蝇寄生对不同微生境中紫茎泽兰花序及种子数量的影响

微生境		花序数 $\times 10^3/m^2$	单花序种子数 /个	总结实量 $\times 10^3/m^2$	寄生率/%
沟谷地	未寄生	5.23±0.35 a	71.01±8.36 a	371.67±15.25 a	31.7±3.4 a
	寄生	3.88±0.27 b	64.16±2.57 ab	249.23±13.61 cd	
撂荒地	未寄生	3.31±0.39 c	61.25±2.30 b	202.79±18.58 de	13.5±1.1 c
	寄生	2.61±0.31 d	57.21±2.65 b	149.24±14.65 f	
果园	未寄生	4.25±0.67 ab	73.53±5.35 a	312.53±23.05 b	26.1±4.5 b
	寄生	3.28±0.45 c	65.28±7.62 ab	213.64±19.57 d	
林地	未寄生	3.93±0.57 b	67.38±10.12 ab	264.59±26.30 c	28.1±2.3 ab
	寄生	2.96±0.28 c	60.26±6.05 b	178.47±14.46 e	

注:表中同一列数据后标字母不同表示差异显著($P<0.05$)。

2.2 不同微生境中泽兰实蝇寄生对紫茎泽兰种子特性的影响

(1)种子长度:紫茎泽兰种子的平均长度在4种微生境间存在一定的差异(表2)。未被寄生时沟谷地与果园的种子较长,撂荒地最短。寄生后,沟谷地和果园生境中的种子长度显著缩短,其中沟谷区的减幅最大,为14.9%,撂荒地和林地种子长度变化较小。

(2)种子宽度:泽兰实蝇寄生对紫茎泽兰种子平均宽度有明显的影响(表2),在各微生境中表现出的差异同其种子长度的变化基本一致。除撂荒地外其余生境内的种子宽度均较未被寄生的显著下降。

(3) 种子千粒重:不同生境中由于水肥条件等差异,紫茎泽兰种子的平均千粒重存在一定差异,但这种差异不显著;而泽兰实蝇寄生对种子千粒重影响很大,在4个生境内均显著下降,其中在林地的降幅最大(表2)。

(4) 种子活力:紫茎泽兰种子的产量较高,其种子活力亦较高,活力水平平均在57.3%~73.7%之间(表2)。泽兰实蝇寄生对种子活力影响显著,在各生境内下降均达显著水平,平均降幅在29.8%~37.9%之间。

表2 泽兰实蝇寄生对不同微生境中紫茎泽兰种子特性的影响

微生境		种子长度/mm	种子宽度/mm	千粒重/g	活力/%
沟谷地	未寄生	1.94±0.11 a	0.45±0.13 a	0.042±0.002 a	73.7±1.2 a
	寄生	1.65±0.10 b	0.37±0.07 b	0.033±0.001 bc	48.4±0.8 c
撂荒地	未寄生	1.63±0.07 b	0.34±0.05 bc	0.038±0.001 ab	57.3±0.7 b
	寄生	1.48±0.04 b	0.29±0.02 c	0.030±0.001 c	40.2±0.5 d
果园	未寄生	1.89±0.12 a	0.45±0.11 a	0.041±0.001 a	73.0±1.5 a
	寄生	1.61±0.03 b	0.38±0.06 b	0.032±0.001 c	45.3±0.4 cd
林地	未寄生	1.72±0.06 ab	0.38±0.10 b	0.040±0.001 a	66.6±0.8 ab
	寄生	1.58±0.03 b	0.31±0.05 c	0.031±0.001 c	41.4±0.6 d

2.3 不同微生境中泽兰实蝇寄生对紫茎泽兰种子萌发率及出苗率的影响

(1) 种子萌发率:泽兰实蝇寄生后,各微生境中的种子萌发率均显著下降(图1)。各生境中未被寄生的种子平均萌发率为63.30%,明显高于寄生后的42.24%。寄生后种子萌发率下降幅度最大的是撂荒地,其余依次为沟谷地、林地和果园。

(2) 种子出苗率:在4种微生境中,未被泽兰实蝇寄生的种子平均出苗率为41.16%(图1),寄生后降为14.60%。其中寄生后出苗率降幅最大的是采自撂荒地的种子,达69.14%,其余依次为采自林地、沟谷地和果园的种子,分别为66.2%、62.5%和58.7%。可以看出,泽兰实蝇寄生导致种子的饱满度下降,从而导致各生境中种子的出苗率大幅下降。

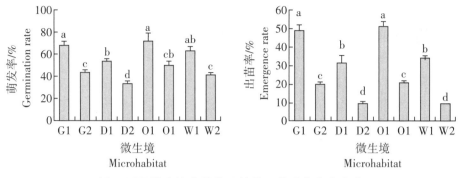

图1 不同微生境中紫茎泽兰种子萌发率和出苗率

注：G为沟谷地，D为撂荒地，O为果园，W为林地；1为未被泽兰实蝇寄生的种子，2为被寄生后的种子；相同字母表示差异不显著（$P \geq 0.05$）。

3 讨论

已有的研究表明，泽兰实蝇在我国西南地区每年发生4~6代，在云南、贵州的寄生率可以高达60%以上。本研究在四川攀西地区的调查发现，春季泽兰实蝇的寄生率最高，但不超过40%，其寄生率差异的原因尚待进一步研究。在不同微生境中，泽兰实蝇的寄生率显著不同，其寄生率的高低与生境中土壤水分含量呈正相关，其原因可能与泽兰实蝇偏好寄生繁茂且幼嫩的枝条，以及与紫茎泽兰的生长分布规律有关。紫茎泽兰是喜湿性植物，在水分充足条件下生长旺盛、茎干幼嫩，而干旱环境下的植株则相反。泽兰实蝇是紫茎泽兰的专性寄生性天敌，对寄主有很强的搜寻能力，紫茎泽兰的生长态势直接影响了泽兰实蝇的分布取向。

泽兰实蝇寄生后对紫茎泽兰结种量的影响显著，平均降幅达31.9%。这可能是因为泽兰实蝇寄生在幼嫩茎枝的端部，蛀食后刺激被害部位形成膨大虫瘿，阻碍了养分向生殖部分的输送，从而更多地转向对营养生长的供给所致。刘光耀等研究表明，泽兰实蝇寄生影响紫茎泽兰地上部分生长及根系发育，导致紫茎泽兰营养不良。本研究分析表明，构成紫茎泽兰种子产量的3个因素即单位面积内的花序数、种子的千粒重和花序内的种子数，寄生后影响最大的是种子的千粒重，其次为花序数，在4个微生境中这两个指标均显著下降，但对每个花序内结籽数影响不明显，这与前人的研究结果相符。

寄生对紫茎泽兰种子产出量的抑制作用表现为与土壤湿度呈正相关，对种子千粒重抑制效应则与生境中光照强度呈负相关，而与土壤养分的相关性不明显。前者是由于随着土壤湿度增加，紫茎泽兰生长愈为茂盛，导致泽兰实蝇寄生率上升所致，总之泽兰实蝇对环境的选择行为还有待进一步研究。

植物种子质量的优劣、萌发率的高低、萌发所需条件及其与微生境的协调性等是影响植物生存及扩散的重要因素。泽兰实蝇寄生导致紫茎泽兰种子的长度和宽度显著下降，即降低了种子的饱满程度，从而抑制了种子的发芽势，种子出土能力大为减弱，所以在一定程度上控制了紫茎泽兰通过有性繁殖进行扩散；此外，寄生对紫茎泽兰种子萌发率和出苗率的影响在撂荒地生境效果更为明显，说明在土壤营养、水分条件较差的生境中，寄生对紫茎泽兰有性繁殖能力形成的抑制作用更为明显。

结合以往的报道可以看出，泽兰实蝇作为紫茎泽兰的专性寄生性天敌，对于紫茎泽兰的扩散蔓延具有较好的控制作用，但就实际情况来看，单纯依赖泽兰实蝇，很难达到有效控制紫茎泽兰的目标。此外，泽兰实蝇作为一种引进的非土著天敌，其控制效果及生态效应仍需进行更为深入的评价。

原文刊载于《植物保护学报》，2006年第4期

温度、土壤含水量和埋蛹深度对柑橘大实蝇羽化的影响

唐松,宫庆涛,豆威,王进军,赵志模

摘要:为揭示环境因子对柑橘大实蝇成虫羽化的影响,采用三元一次正交组合设计分析,分别建立了成虫羽化出土率、存活率、始见日以及羽化历期与温度、土壤含水量和埋蛹深度及其互作关系的回归方程。在4个方程中温度的偏回归系数最高,分别为19.37、22.14、15.88和9.63,且均为负相关;土壤含水量主要影响成虫的羽化出土率和存活率,其偏回归系数分别为6.92和12.02,呈负相关关系;埋蛹深度对成虫羽化出土率和存活率有一定影响,其偏回归系数分别为2.86和1.39,呈正相关关系。因子间互作主要影响成虫的存活率,其中温度×土壤含水量互作是其单因子作用的叠加,而温度×埋蛹深度以及土壤含水量×埋蛹深度的互作在一定程度上抵消了温度和土壤含水量对成虫存活率的影响。

关键词:柑橘大实蝇;温度;土壤含水量;土壤深度;羽化

Effects of Temperature, Soil Humidity and Depth of Buried Pupae on Adult Emergence of *Bactrocera minax*

Tang Song, Gong Qingtao, Dou Wei, Wang Jinjun, Zhao Zhimo

Abstract: With the objective of determining the effects of environmental factors on the adult emergence of *Bactrocera minax*, temperature, soil moisture content, pupae buried depth and the interactions among these three factors were analyzed with respect to the adult emergence characteristics, such as emergence percentage, survival rate, the earliest date of adult appearing, and emergence duration, using orthogonal combination experiment to develop the regression equations of above three factors at different levels. The results showed that among four equations, the partial regression coefficients of temperature were all the highest, which were 19.37, 22.14, 15.88 and 9.63, and the relationships were negative correlated. The adult emergence percentage and survival rate was mainly affected by soil moisture content with the partial regression coefficient of 6.92 and 12.02, and the relationships were also negative correlated. Besides, the positive correlations of the buried depth of pupae with both emergence percentage and survival rate were found with the partial regression coefficient of 2.86 and 1.39. The interactions among these factors mainly affect the adult survival rate. There were additive effects of the interaction between temperature and soil humidity, however, the interaction between temperature×pupae depth and soil humidity×pupae depth were opposite and had antagonistic effects.

Key words: *Bactrocera minax*; temperature; soil humidity; depth; emergence

柑橘大实蝇 *Bactrocera*（*Tetradacus*）*minax*（Enderlein）属双翅目实蝇科寡鬃实蝇亚科,寄主有柑橘属 *Citrus* 的甜橙、酸橙、柚、焦柑、王柑、椪柑、温州蜜橘、红橘、京橘、柠檬、香橼、佛手,金橘属 *Fortunella* 的金橘和枳属 *Poncirus* 的枸橘等。国内主要分布于贵州、四川、重庆、云南、湖南、湖北、广西和陕西等省份,国外分布于南亚次大陆的印度和不丹等国。柑橘大实蝇是柑橘类果树的重要害虫,以幼虫潜居在果实内取食为害,使果内局部或全部成为粉糊状,或造成果实未熟先黄而脱落,其为害损失一般在10%～20%,严重时达50%以上,甚至绝收。柑橘大实蝇1年发生1代,蛹在土壤中历经秋、冬两季约5—7个月的时间,是所有虫态中发育历期最长且受环境因素影响最大的发育阶段。蛹的羽化时间和羽化数量直接决定当年成虫的发生期和发生量。据 Hulthen & Clarke 报道,昆士兰实蝇 *Bactrocera tryoni* 蛹的羽化受土壤结构和土壤含水量的影响。罗禄怡和陈长风认为温度和土壤湿度是影响成虫羽化的主要因子,气温20～25 ℃是蛹发育的最适温区,30 ℃以上蛹不能羽化,30～35 ℃下5～8 d死亡,低温5～10 ℃蛹亦不能羽化;当土壤含水量为15%～20%时,蛹的羽化率最高。但多个因子对成虫羽化的联合影响未见报道,为此,本研究采用三元一次正交组合设计,探讨了温度、土壤含水量和埋蛹深度及其互作对柑橘大实蝇成虫羽化的影响,旨在为该虫的预测预报和制定防治策略提供依据。

1 材料与方法

1.1 供试虫源

2010年10月,在果园捡拾大量蛆果,于室内将蛆果分置于盛有15 cm橘园土的塑料盆中,保持土壤湿度20%左右,待幼虫全部脱果化蛹后,于2011年3月17日即试验前1天,在塑料盆中取蛹作为供试虫源。

1.2 试验设计

以温度(x_1)、土壤含水量(x_2)和埋蛹深度(x_3)为试验因子,按三元一次正交组合设计处理试虫。各因子水平及编码见表1,试验结构矩阵见表2。每个编号处理10头蛹,重复3次。

1.3 处理方法

试验设计中的土壤含水量是指占供试土壤饱和含水量的百分比。在配制不同含水量土壤时,先将土壤置于烘箱中,在130 ℃下烘干至恒重,然后将其分为3份,按试验设计加水,充分混合均匀。试验设计中的埋蛹深度是指蛹位置至土表的距离,在进行不同埋蛹深度处理时,先按试验设计将不同含水量土壤分装在直径约10 cm、高20 cm的塑料杯中,装土深度分别为13、9和5 cm,然后平置试蛹,再加土至15 cm,杯口用保鲜膜拴扎。将埋有试蛹的塑料杯按试验设计置于15、20和25 ℃(±0.5 ℃)的恒温箱中进行处理。

1.4 观察记载

试验处理时间为2011年3月18日。从任何1个处理首次出现成虫起,每天观察1次,记载各编号出土成虫数并区分雌雄,直到连续7 d无成虫出现为止。最后将各杯土壤倒出,观察并记载蛹壳数和死蛹数。羽化特征包括:羽化出土率,指观察到的出土成虫数占供试蛹数的百分比;成虫存活率,指出土成虫数占蛹壳数即实际羽化数的百分比;成虫始见日(也称成虫初见日),指成虫出土最早的日期(按4月21日=1转换);羽化历期,指成虫始见日至终见日历经的天数。

试验数据的方差分析、回归方程的建立和回归系数的t测验均采用 Excel 软件完成。

表1 温度、土壤含水量及埋蛹深度试验水平

试验因子	编码			间距
	-1	0	1	
温度x_1/℃	15	20	25	5
土壤含水量x_2/%	10	30	50	20
埋蛹深度x_3/cm	2	6	10	4

2 结果与分析

2.1 羽化出土情况

供试蛹数为420头,羽化即蛹壳数244头,占总蛹数的58.1%;出土成虫212头,占羽化数的86.9%,占总蛹数的50.5%;在出土成虫中雌性103头、雄性109头,雌雄性比为1∶1.06;成虫始见日为4月21日,终见日为6月16日,羽化历期共57 d。

温度(x_1)、土壤含水量(x_2)和埋蛹深度(x_3)三因子不同水平组合下,柑橘大实蝇的羽化出土率(Y_1)、成虫存活率(Y_2)、成虫始见日(Y_3)和羽化历期(Y_4)见表2。方差分析结果表明,上述4个羽化特征在不同编号间均达极显著差异,因此可以分别建立蛹的4个羽化特征与3个因子及其互作的回归方程,通过分析方程偏回归系数的绝对值大小和正负符号,可以辨明各因子及其互作对4个羽化特征影响的大小和方向(表3)。

表2 试验结构矩阵及结果

试验编号	温度x_1/℃	土壤含水量x_2/%	埋蛹深度x_3/cm	羽化出土率/%	成虫存活率/%	成虫初见日/d	羽化历期/d
1	1	1	1	3.33	20.00	2	1
2	1	1	-1	0.00	0.00	2	0
3	1	-1	1	13.33	66.67	2	2
4	1	-1	-1	6.67	50.00	1	2
5	-1	1	1	50.00	83.33	35	21

续表

试验编号	温度 x_1/℃	土壤含水量 x_2/%	埋蛹深度 x_3/cm	羽化出土率 /%	成虫存活率 /%	成虫初见日 /d	羽化历期 /d
6	-1	1	-1	43.33	92.86	34	18
7	-1	-1	1	73.33	91.67	34	16
8	-1	-1	-1	70.00	100.00	31	27
9	0	0	0	76.67	100.00	5	18
10	0	0	0	70.00	91.30	7	15
11	0	0	0	76.67	92.00	5	12
12	0	0	0	80.00	92.31	5	11
13	0	0	0	66.67	83.33	5	22
14	0	0	0	76.67	85.19	4	15

注：成虫始见日以"4月21日=1"类推。

表3　柑橘大实蝇羽化特征组间的 F 检验

羽化特征	自由度(DF)	均方(MS)	F	P
羽化出土率(Y_1)	13	2 906.96	11.96	3.41E-08
成虫存活率(Y_2)	13	3 019.42	6.69	1.35E-05
羽化初见日(Y_3)	13	755.03	161.79	9.18E-23
羽化历期(Y_4)	13	75.61	5.92	4.12E-05

注：试验重复3次。

2.2 三因子对成虫羽化出土率的影响

根据表2的资料，对柑橘大实蝇的羽化出土率(Y_1)进行反正弦转换后，用温度(x_1)、土壤含水量(x_2)和埋蛹深度(x_3)3个单因子及其互作对羽化出土率(Y_1)进行三元一次正交组合回归分析，其回归方程未达显著水平（F=1.414 9，P=0.327 9）；而仅以3个单因子与羽化出土率进行回归分析，其建立的回归方程①达到显著水平（F=3.99；P=0.041 6）。这说明因子间的互作对羽化出土率的影响被各个单因子的作用所抵消。

$$Y_1=43.36-19.37x_1-6.92x_2+2.86x_3\ (r=0.738\ 1) \quad \text{①}$$

由于该试验具有正交性,因而对建立的回归方程可以固定2个因子的水平为零而得到另一个因子与羽化出土率的回归关系。从回归方程①的偏回归系数绝对值看,温度对羽化出土率的影响最为显著,偏回归系数为19.37;土壤含水量的影响次之,偏回归系数为6.92;埋蛹深度的影响最小,偏回归系数为2.86。从回归系数的符号看,温度和土壤含水量与羽化出土率呈负相关,表明在试验设计的水平下,随着温度和土壤含水量的提高,成虫的羽化出土率相应降低,而埋蛹深度与羽化出土率呈正相关,表明随着埋蛹深度的增加,羽化出土率相应提高。

2.3 三因子及其互作对成虫存活率的影响

根据表2的资料,对柑橘大实蝇成虫存活率(Y_2)进行反正弦转换后,对温度(x_1)、土壤含水量(x_2)和埋蛹深度(x_3)3个因子及其互作与成虫存活率(Y_2)进行三元一次正交组合回归分析,建立的回归模型②达显著水平($F=3.913\ 1$;$P=0.048\ 59$)。

$$Y_2=62.38-22.14x_1-12.02x_2+1.39x_3-6.27x_1x_2+7.68x_1x_3+3.15x_2x_3\ (r=0.877\ 7) \quad \text{②}$$

从模型②的偏回归系数绝对值看,单因子中温度对柑橘大实蝇成虫存活率的影响最大,偏回归系数为22.14;土壤含水量的影响次之,偏回归系数为12.02,埋蛹深度的影响最小,偏回归系数仅为1.39。两因子交互作用的影响以温度×埋蛹深度最大,偏回归系数为7.68;温度×土壤含水量次之,偏回归系数为6.27;土壤含水量×埋蛹深度的影响最小,偏回归系数为3.15。从偏回归系数的符号看,3个单因子与成虫存活率的关系和其与羽化出土率的关系完全相同,即温度和土壤含水量与成虫存活率呈负相关,与埋土深度呈正相关;在互作因子中,温度×土壤含水量与成虫存活率呈负相关,说明这2个单因子互作对其单因子的作用有叠加效应;而温度×埋蛹深度和土壤含水量×埋蛹深度与成虫存活率呈正相关,说明这2个因子的互作在一定程度上抵消了温度和土壤含水量对成虫存活率的影响。

2.4 三因子及其互作对成虫始见日及羽化历期的影响

根据表2的资料,对成虫始见日(Y_3)和羽化历期(Y_4)与温度(x_1)、土壤含水量(x_2)和埋土深度(x_3)及其互作进行三元一次正交组合回归分析,分别得到回归方程③和回归方程④。经F检验,2个回归方程均达到了显著水平(方程③:$F=4.4002$,$P=0.037$;方程④:$F=5.2494$,$P=0.0234$)。

$Y_3=12.29-15.88x_1+0.63x_2+0.63x_3-0.38x_1x_2-0.38x_1x_3-0.38x_2x_3$($r=0.8891$) ···③

$Y_4=12.86-9.63x_1-0.88x_2-0.88x_3+0.13x_1x_2+1.13x_1x_3+1.88x_2x_3$($r=0.8823$) ······④

从方程③的偏回归系数绝对值看,在3个单因子中,温度对柑橘大实蝇成虫始见日的影响最为明显,偏回归系数为15.88,土壤含水量和埋蛹深度对成虫初见日的影响极小,两者的偏回归系数均为0.63,未达显著水平。3对互作对成虫初见日的影响不显著,其回归系数均为0.38。从方程③回归系数的符号看,3个单因子中,温度与成虫始见日呈负相关,即温度越高,成虫始见日越早;土壤含水量和埋蛹深度与成虫初见日均呈正相关,即随着土壤含水量和埋蛹深度的增加,成虫始见日相应延迟。3对互作因子与成虫始见日均呈负相关,但其影响均不明显。

从方程④偏回归系数的绝对值看,在单因子中,温度对成虫羽化历期的影响最为明显,偏回归系数为9.63,即温度越高,羽化历期越短,而土壤含水量和埋蛹深度对成虫羽化历期没有显著影响,偏回归系数均仅为0.88;3对互作对成虫羽化历期的影响均不显著,偏回归系数在0.13~1.88之间。从回归系数的符号看,单因子温度、土壤含水量和埋蛹深度与羽化历期均呈负相关,即随着温度升高、土壤含水量和埋蛹深度增加,羽化历期相应缩短;3对互作因子与羽化历期均呈正相关,这与方程③反映的3对互作因子与成虫始见日均呈负相关的情况相反。

3 讨论

本研究应用三元一次正交组合设计,探讨了温度、土壤含水量和埋蛹深度对柑橘大实蝇成虫羽化特征的影响。三元一次正交组合试验设计,不仅减少了试验因子的组合数,而且可应用回归分析方法研究3个单因子及其互作与成虫羽化特征之间的关系。

试验设计的温度、土壤含水量和埋蛹深度是影响柑橘大实蝇成虫羽化特征的主要因子。获取的4个羽化特征中,羽化出土率和成虫存活率直接影响柑橘大实蝇成虫在春季的发生量,而成虫始见日和羽化历期直接影响成虫的发生期。

本研究拟合的4个回归方程中,温度的偏回归系数绝对值均最大,经t测验全部达到显著或极显著水平,表明温度是影响柑橘大实蝇成虫羽化最主要的因子。回归方程中温度的偏回归系数均为负值,即在15~25 ℃条件下,随着温度的升高,其始见日提前,羽化历期缩短,这合乎温度与昆虫生长发育的一般规律,而随着温度升高,成虫羽化出土率和存活率相应降低,与已有的研究报道基本一致,如吕志藻等、易继平报道,柑橘大实蝇蛹在24 ℃以上羽化率下降,甚至不能羽化。

土壤含水量也是影响柑橘大实蝇成虫羽化的重要因素。本研究结果表明,土壤含水量对成虫始见日和羽化历期的影响不明显,而对成虫羽化出土率和存活率的影响较为明显,并均呈负相关,即随着土壤含水量的增加,柑橘大实蝇成虫的羽化出土率和存活率相应降低,这一结论与罗禄怡和陈长凤的羽化最适土壤含水量为15%~20%的结果相一致。土壤含水量过高造成羽化出土率和存活率降低,主要是由于蛹的霉变和腐烂。本研究中蛹的总死亡率达41.9%,已羽化的成虫在土壤中的总死亡率为13.1%,并多分布在土壤含水量高的试验组合中,其死亡原因也是由于蛹的霉变和腐烂。

本研究设计中的埋蛹深度是模拟幼虫在土壤中的化蛹深度。据报道,蛆果中的幼虫发育到3龄后脱果,在土壤3~13 cm深处化蛹。本研究表明,埋蛹深度仅对成虫羽化出土率和存活率有一定影响,即在试验设计水平下,随着埋蛹深度的增加,成虫羽化出土率和成虫存活率有所提高。这可能与本研究使用的砂性土壤和较小的试验空间有关。

研究结果表明,因子互作对成虫羽化出土率、成虫始见日和羽化历期的影响很小,而对成虫存活率的影响较大,3对互作因子的偏回归系数分别为7.68、6.27和3.15。温度和土壤含水量2个单因子的偏回归系数为负值,其互作也为负值,说明这对互作因子对其单因子的作用具有叠加效应。而温度×埋蛹深度和土壤含水量×埋蛹深度2个互作中,温度和土壤含水量2

个单因子的偏回归系数为负值,2个互作的回归系数均为正值,这说明这2对互作因子在一定程度上抵消了温度和土壤含水量对成虫存活率的影响。

本研究阐明了温度、土壤含水量和化蛹深度及其互作对柑橘大实蝇羽化的影响,为成虫发生期和发生量预测提供了依据。但由于本研究在实验室条件下进行,其全部因子在3个水平上都是恒定的,因此研究结果的应用应与自然条件下的观察结合起来。

原文刊载于《植物保护学报》,2012年第2期

柑橘大实蝇羽化出土及橘园成虫诱集动态研究

宫庆涛,武可明,唐松,何林,赵志模

摘要:柑橘大实蝇是柑橘类果树上的重要害虫。预测该虫的羽化出土进度、掌握成虫发生动态是指导橘园成虫期防治的重要依据。本研究通过在25 ℃恒温、室内常温和室外网室3种条件下饲养柑橘大实蝇的蛹,以逐日观察成虫羽化出土数量;在重庆武隆、四川江油等5个地区共设置240个麦克菲尔(McPhail)诱集器,以糖酒醋液和水解蛋白为诱饵诱集成虫,得到柑橘大实蝇成虫羽化出土的逐日数量和橘园成虫诱集的逐期数量。用逻辑斯蒂模型拟合成虫羽化出土和橘园成虫诱集动态,结果表明,成虫的始盛期、高峰期和盛末期在25 ℃恒温条件下分别为4月25日、28日和30日,盛期的持续时间为6 d;在室内常温条件下分别为5月3日、7日和10日,盛期的持续时间为8 d;在室外网室条件下分别为5月8日、14日和18日,盛期的持续时间为11 d;橘园诱集成虫分别为6月2日、14日和26日,盛期的持续时间为25 d。随着羽化期温度的提高,柑橘大实蝇羽化出土期提前,历期缩短,羽化整齐。虽然网室成虫羽化和橘园成虫诱集都处于室外条件,但后者的始盛期、高峰期和盛末期比前者分别迟了25、31和39 d。因此,建议采用室外网室饲养蛹的方法监测柑橘大实蝇成虫的发生期,若仅凭橘园诱集成虫的数据,因其滞后性十分明显,对指导柑橘大实蝇成虫防治的意义不大。

关键词:柑橘大实蝇;羽化出土动态;橘园成虫诱集动态

Emergence Dynamics of Adults of the Chinese Citrus Fly *Bactrocera minax*

Gong Qingtao, Wu Keming, Tang Song, He Lin, Zhao Zhimo

Abstract: The Chinese citrus fly [*Bactrocera minax* (Enderlein)] is one of major citrus pests in China. The forecast and understanding of emergence progress and seasonal dynamics of adult flies is the important guidance for controlling of this insect pest. The emergence of the adults was observed at the constant temperature 25 ℃, indoor room temperature, and an outdoor screening house, using the fly pupal rearing method. The dynamics of adults in nature was followed in the citrus orchards by setting up 240 McPhail traps with sugar-alcohol-vinegar solution and proteinaceous lures at five locations of Chongqing and Sichuan. At the constant temperature 25 ℃, adults started to emerge on 25 April, and peaked on 28 April. The last adult emerged on 30 April, and successive time of the peak period was 6 days; at the condition of indoor room temperature, they were 3th, 7th and 10th of May, respectively, and the successive time was 8 days; in the outdoor screening house, adults started to emerge on 8 May, peaked on 14 May and ended on 18 May, and successive time was 11 days; trapping in citrus orchards yielded the first adults on 2 June, with a peak of 14, and end of 26 June, and the successive time was 25 days. The peak period of the adult eclosion unearthed was shifted to an earlier date, and the successive time was decreased under increased temperatures. Although pupae rearing in the screening house and adult trapping all were under natural, outdoor conditions, the beginning, peak, and end dates for trapped adults were later than in the screening house by 25 days, 31 days and 39 days, respectively. Therefore, guidance on the control for adult fly should not solely be based on citrus orchard trapping data due to its obvious time-lag. Pupal rearing in screening houses could better forecast the emergence of adult Chinese citrus fly.

Key words: Chinese citrus fly; dynamics of adult emergence; dynamics of adult trapping in the citrus orchard

柑橘大实蝇 *Bactrocera minax* (Enderlein)属双翅目实蝇科寡鬃实蝇亚科,国内主要分布于贵州、四川、重庆、云南、湖南、湖北、河南、广西和陕西等省,为害甜橙、酸橙、柚子、温州蜜橘、红橘、京橘、葡萄柚和佛手等柑橘类植物的果实,为害历史已有百余年之久。柑橘大实蝇每年发生1代,成虫产卵于果实表皮层中,幼虫潜居果内为害,老熟幼虫脱果入土化蛹、越冬。柑橘大实蝇是柑橘类果树的主要害虫,造成的损失一般在10%~20%,严重时达60%~90%。以往,各地区多采用果园成虫诱集预测柑橘大实蝇成虫的发生期,并以此作为指导田间药剂防治的依据,但这种方法预测的发生期明显滞后。本研究在25 ℃恒温、室内常温和室外网室3种条件下,观察了柑橘大实蝇羽化出土动态,并与果园成虫诱集动态进行比较,这对于确切掌握柑橘大实蝇成虫发生期和指导田间药剂防治具有重要意义。

1 材料与方法

1.1 试验材料

2010年10月,在四川省江油市果园捡拾大量蛆果备用。

1.2 试验方法

(1)2010年10月,将采集到的蛆果置于25个分别盛有5 cm厚沙土的塑料盆(25 cm×40 cm)中,每盆均匀搁置蛆果约20个,待其化蛹后,于2011年3月19日将其中4盆放入(25±0.5)℃恒温下,其余20盆继续放在室内常温下,逐日观察羽化出土的雌、雄成虫数。

(2)2010年10月,将约800个蛆果均匀置于室外网室内装有砂壤土的水泥槽(120 cm×300 cm)中,逐日观察羽化出土的雌、雄成虫数。

(3)2011年4月中旬至9月上旬,在四川省江油市,重庆市万州区、武隆县、云阳县、巫山县5个地区,各选择1个往年柑橘大实蝇发生较重的果园,在约1 300 m²的果园中,各挂置48个麦克菲尔(McPhail)诱集罐(每间隔1株树挂1个罐),利用5%糖醋酒液和水解蛋白液等进行诱集,每7 d观察1次,记录诱集到的雌、雄成虫数。

1.3 数据处理与分析

将各处理的成虫数和橘园诱集数进行逐日或逐期累加,以各日或各期累计数占总数的百分比作为成虫羽化出土进度或成虫诱集进度,并分别拟合逻辑斯蒂方程。方程拟合、做曲线图、方差分析和性比的 t 测验均利用Excel软件完成。

2 结果与分析

2.1 柑橘大实蝇成虫羽化出土动态

图1给出了25 ℃恒温、室内常温和室外网室条件下柑橘大实蝇逐日羽化出土动态,表1给出了上述3种条件下柑橘大实蝇出土成虫总数、雌性

比、成虫始见日、终见日和总历期。

从图1和表1可以看出,25 ℃恒温、室内常温和室外网室3个处理的出土成虫数分别为115、594和710头(成虫羽化出土的数量主要受各处理供试虫量的影响,本研究未统计幼虫的化蛹率和蛹的羽化出土率)。雌性比分别为53.91%、55.72%和50.70%;t测验表明,除室内常温下出土成虫的雌性比较高,不符合大多数生物1∶1的规律外($t=2.808\,5>t_{0.05,593}=1.964\,0$),其余2个处理的性比均接近1∶1(25 ℃下:$t=0.748\,3<t_{0.05,114}=1.980\,1$;室外网室:$t=0.337\,8<t_{0.05,709}=1.963\,3$)。从成虫出土的初见日、高峰日和终见日来看,25 ℃恒温下最早,室外网室最晚,室内常温下居两者之间。但3种处理的初见日、高峰日和终见日提前或延迟的时间并不相等。例如,室外网室的初见日较室内常温下延迟了13 d,而高峰日和终见日仅分别延迟4 d和10 d;室内常温的初见日较25 ℃恒温下延迟6 d,而高峰日和终见日分别延迟5 d和14 d。这种差异主要是由于3种处理出土成虫的逐日分布和羽化总历期不同造成的。在25 ℃恒温条件下,出土成虫的逐日数量分布大体上呈右偏态分布,羽化总历期最短,仅为13 d;室内常温下的逐日数量分布基本上呈正态分布,羽化总历期最长,为21 d;而室外网室条件下的逐日数量分布呈明显的左偏态分布,羽化总历期为18 d。

表1　柑橘大实蝇成虫羽化出土的生物学参数

处理条件	成虫总数/头	雌性比/%	初见日/(m-d)	高峰日/(m-d)	终见日/(m-d)	出土历期/d
25 ℃恒温	115	53.91	04-22	05-02	05-04	13
室内常温	594	55.72	04-28	05-07	05-18	21
室外网室	710	50.70	05-11	05-11	05-28	18

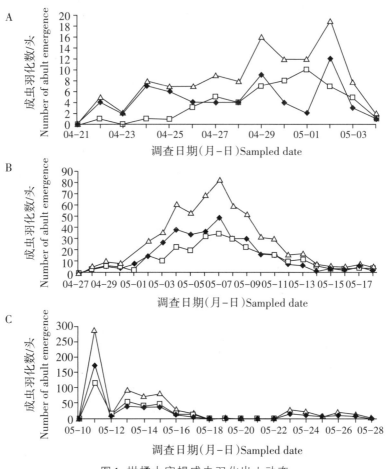

图1 柑橘大实蝇成虫羽化出土动态
(A:25 ℃恒温;B:室内常温;C:室外网室)

2.2 柑橘大实蝇成虫橘园诱集动态

2011年在江油、武隆等5个地区的柑橘园共诱集柑橘大实蝇成虫628头。其中,雌虫361头,雄虫267头,雌性占57.50%。在柑橘园诱集到成虫的初见日为5月4日,终见日为8月6日,诱集到成虫的时间长达94 d。但5月18日之前,仅诱集到2头雄虫,从5月18日开始诱集量急剧增多,高峰日出现在6月1日,当日诱集量占总诱集量的37.30%(图2)。橘园成虫诱集动态与室外网室的大体一致,均呈左偏态分布。

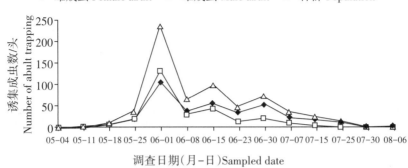

图2 柑橘大实蝇成虫橘园诱集动态

2.3 柑橘大实蝇成虫羽化出土与橘园成虫诱集的动态模型

为了得到柑橘大实蝇成虫羽化和果园成虫诱集进度为16%、50%和84%的始盛期、高峰期和盛末期,将各日或各期累计数换算为羽化进度或诱集进度(Y),并与羽化出土时间或诱集时间(X)拟合为逻辑斯蒂模型。在方程拟合时,均令其成虫初见日=1。各处理羽化出土或诱集进度的实际值和相应模型的拟合值曲线见图3。

25 ℃恒温:$Y_1 = \dfrac{100}{1+e^{4.5535-0.7005X}}$ ($R^2=0.8424, F=58.807, P=9.74\text{E}-06$)

室内常温:$Y_2 = \dfrac{100}{1+e^{4.7116-0.4870X}}$ ($R^2=0.9763, F=781.730, P=6.71\text{E}-17$)

室外网室:$Y_3 = \dfrac{100}{1+e^{0.8339-0.3066X}}$ ($R^2=0.7731, F=54.522, P=1.54\text{E}-06$)

果园诱集:$Y_4 = \dfrac{100}{1+e^{5.8009-0.1386X}}$ ($R^2=0.9694, F=380.414, P=1.87\text{E}-10$)

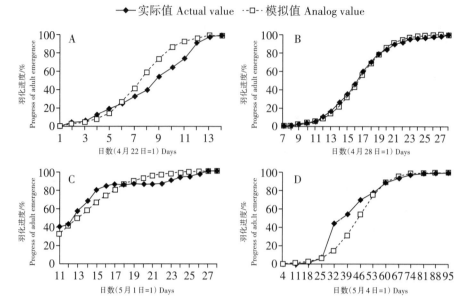

图3 柑橘大实蝇成虫羽化和橘园诱集动态模型

注：A.25 ℃恒温；B.室内常温；C.室外网室；D.果园诱集。

根据动态模型计算,得到4种处理下成虫羽化或诱集的始盛期、高峰期和盛末期(见表2)。25 ℃恒温条件下,柑橘大实蝇成虫羽化出土的始盛期在4月25日,较室内常温、室外网室和果园诱集分别早8、13和38 d;25 ℃恒温下的高峰期在4月28日,较室内常温、室外网室和果园诱集分别早9、16和47 d;25 ℃恒温的盛末期在4月30日,较室内常温、室外网室和果园诱集的分别早10、18和57 d。从表2还可看出,25 ℃恒温、室内常温和室外网室成虫羽化出土盛期(羽化16%～84%)的时间较短,仅为6～11 d;而橘园成虫诱集盛期的时间较长,为24 d。

表2 柑橘大实蝇羽化出土和果园诱集的始盛期、高峰期和盛末期

处理条件	始盛期/(m-d)	高峰期/(m-d)	盛末期/(m-d)	盛期历期/d
25 ℃恒温	04-25	04-28	04-30	6
室内常温	05-03	05-07	05-10	8
室外网室	05-08	05-14	05-18	11
果园诱集	06-02	06-14	06-26	24

3 结论与讨论

在25 ℃恒温、室内常温和室外网室3种条件下,对柑橘大实蝇羽化出土数量的逐日观察表明,温度是影响其羽化出土进度的主要因素。这一结论与罗禄怡和陈长风、王小蕾和张润杰、方正茂的报道相同。本研究于2011年3月19日至5月28日进行,试验期间的室外平均温度为18.9 ℃,试验地区春季室内温度一般高于室外3~4 ℃,则室内均温在22~23 ℃,低于25 ℃的恒温处理。从表1、2可知,25 ℃恒温条件下,柑橘大实蝇羽化出土初见日、高峰日和终见日以及出土盛期均早于室内常温和室外网室,而室内常温又早于室外网室。3种条件下柑橘大实蝇成虫羽化出土的总历期和盛期以25 ℃恒温最短,分别为13 d和6 d;室内常温次之,分别为21 d和8 d;室外网室最长,分别为18 d和11 d。由此可见,随着温度的升高,柑橘大实蝇的羽化出土期提前,历期缩短,羽化整齐;而随温度的降低,羽化出土期推迟,历期延长,羽化趋向于分散。此外,羽化期的早迟还与温差大小有关。据刘治才等报道,平坝地区温差小,柑橘大实蝇羽化出土要比温差大的山区早。这也可能是25 ℃恒温下,柑橘大实蝇羽化出土早于室内常温,更早于温差变化较大的室外网室的原因之一。

通过对室外网室柑橘大实蝇逐日羽化出土数量的观察发现,除温度因素外,土壤湿度(降雨量)也是影响柑橘大实蝇羽化的一个重要因素。在本研究期间,当年春旱严重,从5月3日至9日,降雨量累计仅0.2 mm,羽化量为0头;5月10日降雨量达19.2 mm,导致5月11日柑橘大实蝇单日羽化出土量达285头,占成虫总出土量的40.10%。王小蕾和张润杰、张佳峰、吕志藻和赵逸潮认为,冬季雨量适中,土壤含水量在10%~15%,有利于蛹的越冬和羽化,过干或过湿都会影响柑橘大实蝇的羽化率及羽化时期;王志静等通过在湖北荆门连续7年的观察发现,气温低,降雨少,越冬蛹羽化期推迟;而气温较高,降雨适量,则羽化期提前。罗兴中等在湖北丹江口地区发现,3—4月阴雨连绵,土壤湿度大,有利于成虫羽化。鲁红学等也认为,当年4、5月份降雨时间和雨量直接关系到越冬蛹的羽化初见日、羽化持续时间和羽化存活率。本研究观察的结果与这些报道一致。

本研究通过比较大田成虫诱集与室外网室羽化动态发现,虽然两者都处于室外条件,但前者的始盛期、高峰期、盛末期比后者分别晚了25、31和39 d。其原因除了诱集成虫的柑橘园多在丘陵坡地,温度和土壤湿度较室外网室低,土质也较黏重外,还可能与诱集剂(食物诱饵)的诱集能力有关。本试验用5%的糖酒醋液和水解蛋白作为诱饵,240个诱集罐共诱到柑橘大实蝇成虫628头,平均每罐仅2.62头,绝大多数诱集到的成虫是在蛹羽化盛期之后。另外,成虫出土后具有迁入果园附近的杂木林中取食补充营养,待性成熟后才大量迁回柑橘园产卵的习性,这可能是大田成虫诱集进度比室外网室成虫羽化出土较晚的重要原因。因此,笔者认为,仅以柑橘园诱集成虫的数据预测柑橘大实蝇成虫的发生期,其滞后性十分明显,对指导柑橘大实蝇成虫药剂防治的意义不大,建议采用室内、室外饲养蛹的方法,观察柑橘大实蝇成虫的羽化出土进度,并综合考虑温度、雨量等因素预测当年成虫发生期,用以指导成虫药剂防治。

原文刊载于《生物安全学报》,2012年第2期

酸雨对植物—害虫—天敌系统的作用

张建萍,王进军,赵志模

摘要:综述了酸雨对植物—害虫—天敌系统的影响和酸雨对害虫的作用机制,探讨了模拟酸雨与自然酸雨对生态系统的不同作用,并展望了进一步研究和发展的方向。

关键词:酸雨;植物;害虫;天敌;生态系统;作用机制

Effect of Acid Rain on a Plant-Insect Pest-Natural Enemy Ecosystem

Zhang Jianping, Wang Jinjun, Zhao Zhimo

Abstract: The effect of acid rain on a plant-insect pest-natural enemy ecosystem is described. The impacts of acid rain on insect include direct and indirect effect. Growth, development, survivorship, reproduction and dynamics of insect may be affected not only by acid rain applied to insect directly but also by acid rain acted on plant, soil or natural enemy. By altering host-plant morphological structure, pH value and tissue concentrations of nutrient and defensive compounds, acid rain influences host susceptibility and suitability for insect herbivores. So far, many issues still need to be studied further.

Key words: acid rain; plant; insect pest; natural enemy; ecosystem; mechanism

随着现代工业的迅速发展、人口剧增和城市化趋向,硫化物和氮氧化物被排放得愈来愈多,这些气态化合物在大气中反应生成硫酸和硝酸,随雨、雪、雾等从大气层降落形成酸雨。20世纪70年代以来,欧洲、北美洲和日本等发达地区先后开展了大规模的酸雨研究。我国自1979年开始酸雨的研究和监测,发现长江以北地区的一些城市有酸雨沉降,其中西南地区的酸雨最为严重。酸雨对陆生生物、水生生物、建筑材料、文物和人体健康带来了明显的影响,特别是对森林、湖泊和农业生态系统的影响已引起国内外的关注。

酸雨(pH小于5.6的酸性沉降,包括雨、雪、雹、雾等)作为一种非生物的环境胁迫因素,也会像温度、湿度、光照等一样,可能改变某些植物的理化性质,降低植物的抗性和生态系统的多样性与稳定性,从而引起某些有害生物猖獗发生,由此而造成的间接经济损失可能比酸雨污染本身对农林业的直接危害还要大。在长期协同进化过程中,昆虫(螨)和它们的寄主植物、天敌原处于一种敏感的动态平衡状态,气候变化和人类活动等许多环境因素都会打破这种平衡关系。探讨与研究酸雨对植物—害虫—天敌系统的影响,深化对环境胁迫下植物、害虫、天敌生命活动规律的认识,对正确指导酸雨污染区的害虫防治、天敌保护具有重要意义。

1 酸雨对害虫的直接影响

酸雨对昆虫的直接影响是指酸雨直接作用于虫体而影响昆虫的生长、发育、存活、繁殖和种群动态。有关这方面的研究几乎都是在模拟酸雨或酸雾条件下进行的。Redak等发现以pH为2.75的酸雾直接作用于叶甲(*Trirhabda geminata* Horn)时,在最初几天,延缓了该虫的生长发育和降低了存活率,但一段时间后,酸雾对该虫的存活率及生长发育不起作用,说明随时间增加,叶甲对酸雾胁迫的适应和忍受能力逐渐增强,这与Trumble和Paine早期用模拟酸雾处理金甲豆和菊科寄主植物上的粉纹夜蛾与叶甲所得的结果基本一致。Zhao等在研究酸雨对朱砂叶螨[*Tetranychus cinnabarinus*(Boisduval)]的直接作用中发现,在pH为5.6~3.0的酸度范围内,酸雨对朱砂叶螨虽然没有急性致死作用,但随着酸雨酸度的增加,除卵历期没有大的变化外,其幼螨期、若螨期和产卵前期有所延长,雌成螨寿命和全世代历期缩短;酸雨对该螨种群的生命参数也有明显影响,表现为净生殖率(R_0)、内禀增长率(r_m)、周限增长率(λ)、世代平均周期(T)明显降低,而种群翻倍时间(t)明显延长,说明在pH为5.6~3.0的酸雨直接作用下,朱砂叶螨的生长、发育和繁殖均受到不利的影响。

酸雨对昆虫直接影响的大小和方向与昆虫对酸雨的忍受力及适应能力有关。例如pH小于3的酸雨仅影响叶甲(*Trirhabda geminata*)的生长发育却不致死。而对蜜蜂,当酸雨pH为4.2~3.0时,蜜蜂泌蜜量下降60%,蜂群死亡增多;pH小于3时,泌蜜停止,蜂群惨死。这种情况与其他空气污染对昆虫的直接作用相似,例如SO_2可延长黑尾果蝇(*Drosophila melanogaster*)的生长发育,而对寄生性麦蛾茧蜂(*Habrobracon hebetor* Say)的生长发育没有影响,但能刺激美洲脊胸长蝽[*Oncopeltus fasciatus*(Dallas)]的生长发育和繁殖。

2 酸雨对害虫的间接影响

酸雨对害虫的间接影响是指酸雨通过改变寄主植物的营养、生理代谢、抗性以及改变生物群落中某些成员之间特别是昆虫与其天敌之间的关系而影响害虫。在通常情况下,酸雨对植物并不造成明显的可见伤害,却可以使一些植物的生理特性、形态结构及抗性发生变化,从而改变害虫与植物的相互关系,影响到害虫的生长发育与繁殖,或引起不同害虫的消长与演替。Stinner等的研究表明,酸雨对小地老虎幼虫的取食具有胁迫作用,用高酸度(pH为2.8和4.2)模拟酸雨处理过的玉米饲喂小地老虎(*Agrotis ypsilon*)幼虫,其取食量显著高于常规雨水(pH为5.6)处理的取食量,且幼虫长得较大,发育较快,前蛹期缩短,但同化系数降低。牟树森等报道重庆近郊酸雨严重污染区的蔬菜病虫害增加。朱砂叶螨在pH为4.0酸雨作用的菜豆上平均每叶螨量和危害程度均大于对照。

酸雨对森林生态系统中昆虫的影响研究较多。我国西南地区,酸雨污染严重,对森林生态系统有很大危害。重庆南山马尾松林由于空气污染和酸雨、酸雾的长期作用,使敏感的天敌昆虫逐渐减少,而对污染具有抗性的介壳虫却得到了繁衍的机会,另外一些次生性的蛀干害虫大量攻击因污染而衰弱或濒临死亡的树木,从而使重庆南山马尾松林大面积成片枯死。杨金宽等认为,长期的酸雨污染,致使树木长势衰弱,从而为次生性害虫和病原菌的入侵、定居和繁殖创造了条件。马尾松齿小蠹 *Ips* sp.和松天牛(*Monochamus alternatus* Hope)的发育与酸雨对树木的危害也有一定关系,且酸雨酸度越大,马尾松受害越突出,虫害发生也越严重。

酸雨还可引起湖泊酸化,从而影响水生昆虫的物种多样性。各类水生昆虫对水体酸化的忍耐能力有很大差异。襀翅目、半翅目和鞘翅目昆虫能忍受的最低pH中值为6.0;双翅目能忍受的最低pH中值为5.5;毛翅目昆虫较为耐酸,可发生在pH为5.0以下的水体中。pH大于4.5的水体对背摇蚊(*Chironomus dorsalis* Meigen)的生长发育没有影响;但当pH小于4.5时,它们不能构筑巢筒,提前化蛹,羽化中途停顿,半数死于蛹室之内。

在土壤生态系统中,有关节肢动物(主要是昆虫)对酸雨的敏感度有较多的研究。酸雨中可能包含了H^+,SO_4^{2-},NO_3^-,Cl^-,NH_4^+,Ca^{2+},Na^+,K^+,Mg^{2+}等离子态的物质,不同地区的酸雨,所含离子的种类和离子的浓度有所不同。土壤受酸雨长期作用后,其理化性质和物理结构发生变化,这常常是土壤中小型节肢动物(主要是弹尾目昆虫和螨类)数量增加,而大型节肢动物数量减少的重要原因。Heneghan和Bolger认为,酸雨引起土壤中昆虫数量的变化与酸雨中所含H^+的浓度并无直接关系,而与这些离子形成的化合物关系密切。他们发现在富含NH_4NO_3,H_2SO_4的土壤中,螨类和弹尾目昆虫的数量比富含HNO_3,$(NH_4)_2SO_4$,NH_4Cl的土壤多。

酸雨对植物—害虫系统的影响是动态的。Sabine和Walter通过对菜豆(*Phaseolus vulgaris*)及欧洲山毛榉(*Fagus sylvatica*)上的2种蚜虫豆卫矛蚜(*Aphis fabae scopoli*)和山毛榉叶蚜(*Phyllaphis fagi*)进行模拟酸雾作用试验,发现在喷雾的最初几周,蚜虫的生长受到很大抑制,以后数量逐渐回升。在试验时期内,用酸雾喷过的植株上的蚜虫数始终低于对照植株上的蚜虫数量;他们还用不同pH的人工饲料喂养蚜虫,低pH饲养的第1代蚜虫生长受到抑制,但对第2代蚜虫似乎没有多大影响。

有关酸雨对害虫天敌的影响研究较少。Gunnarsson和Johnsson研究发现,蜘蛛[*Pityohyphantes phrygianus*(Koch)]在pH为4.0酸雨作用下的生长发育与对照没有区别,但在pH为2.2时,该蜘蛛的发育速度显著减慢,且有低的死亡率。一种弗叶甲(*Phratora polaris*)长期取食酸雨刺激的山桦后,使天敌(蚂蚁、步甲和鸟)对它的敏感度增加。另外酸雨还可能增强某些昆虫对病害的抗性,例如松柏锯角叶蜂(*Neodiprion Sertifer* Geoffroy)受酸雨胁迫后,对多角体病毒的敏感性大大降低。

3 酸雨对植物—害虫系统的作用机理

昆虫以特定的植物为其食物是经过长期自然选择的结果。由于酸雨污染的影响,寄主植物的化学成分、防御性化合物和形态结构等可能发生一系列变异,破坏了原已建立的害虫与植物之间的平衡,从而影响到害虫的生长发育和种群动态,并常常使得一些潜在的或次生性的害虫大发生,造成农林植物受害加重。

3.1 酸雨改变植物体内营养物质的质量和数量

当植物生长的环境受到污染时,植物的代谢过程受到影响,代谢产物也发生相应变化。目前大多数学者认为,酸雨污染引起昆虫种群数量变动,主要是由于酸雨改变了寄主植物体内营养物质的质量和数量。例如酸雨能促使菜豆和山毛榉韧皮部氨基酸含量增加,从而促进了豆卫矛蚜和山毛榉叶蚜的生长发育。姬兰柱等认为,酸雨污染引起入侵马尾松的次生性害虫种类和数量显著增加,危害明显加重,其主要原因是马尾松韧皮部多种氨基酸含量上升导致马尾松抗虫性降低。Trumble 等研究发现,当 pH 为 3.0 的酸雾作用于金甲豆(*Phaseolus lunatus* L.)时,叶片总氮含量明显增加,粉纹夜蛾(*Trichoplusia ni* Hübner)对营养物质的消化率提高,生长速度加快,从而改变了该种植物与粉纹夜蛾之间的动态平衡,这是粉纹夜蛾危害加重的重要原因。Paine 等在研究酸雾作用下植物(*Encelia farinosa* A.Gray ex Torr.)与叶甲(*Trirhabda geminata*)的关系时也得到完全相似的结果。也有作者认为,酸雾对植物—害虫动态系统的影响,可能与寄主植物总蛋白和可溶性蛋白含量增加后,改变了对植食性昆虫的敏感度和吸引力有关。

虽然大多数学者认为酸雨通过改变寄主植物体内营养物质的质量和数量而影响昆虫种群数量,但 Redak 却持不同观点,他通过实验证明一种植物(*Encelia farinosa*)叶片的含水量、可溶性蛋白、3 种化学防御物质(farinosin, encecalin 和 euparin)在酸雾作用下与对照的变化规律相似。他认为这些变化是植物本身的代谢变化,而并非酸雾所致。因此认为叶甲(*Trirhabda geminata*)生长速度加快与这些物质变化无关,而可能受其他物质和因素的影响。

3.2 酸雨对植物pH的影响

当植物受到外界环境酸性物质或其他污染物影响时，细胞对pH的同化和缓冲能力就会降低。Wellburn和Wolfender发现，当外界环境酸度降低3个单位，植物细胞液的pH将降低0.3个单位。植物体内pH的变化，常常使昆虫所需营养物质的溶解性和成分发生改变，这可能是酸雨影响害虫取食和生长发育的重要因素。众所周知，蚜虫是一类吸取植物汁液的害虫，它们对食物pH的变化比其他昆虫更为敏感。例如豌豆蚜[*Acyrthosiphon pisum* (Harris)]能够区别pH相差0.3~0.6个单位的食料，它更喜爱在弱酸性的植物部位取食，因此，酸雾和酸雨引起寄主植物pH的变化是促进豌豆蚜生长发育的一个重要原因。

3.3 酸雨改变植物的形态结构

许多昆虫除了以植物为食物外，还以植物作为栖息、产卵或越冬的场所，并受植物所形成的生境小气候的影响。植物的外部形态和内部结构直接影响其所提供的生境的性质，由此影响到昆虫的行为、生长发育和繁殖。例如苹绕实蝇(*Rhagoletis pomonella*)在酸雨(pH<3.8)淋过的寄主果实上产卵量减少，这是因为果实角质层三维结构和表皮化合物特性因高酸度雨水而改变，从而干扰了实蝇在识别和接受寄主时的感觉机制。许多研究表明，当酸雨的酸度超过植物叶片的耐受阈限时，其形态结构发生一系列变化。Evans等研究指出，植物叶片出现可见伤害时的最高pH为3.4，并认为最先受影响的部位是叶面的表皮细胞，随着受酸雨胁迫时间的延长，叶面稍内的细胞层也可能受害。王玮等通过电镜观察发现，青菜经pH为3.0的酸雨处理后，叶的表皮细胞出现轻度瓦解；而pH为2.0的酸雨则使表皮细胞完全瓦解，海绵组织细胞出现部分瓦解，栅栏组织细胞出现不规则排列现象；叶绿体在pH为3.0以下的酸雨胁迫下形态结构发生变化，片层结构被严重破坏，类囊体明显扭曲；在pH为2.0酸雨胁迫下线粒体结构发生明显变化，线粒体脊间增大，内含物减少，酶分子附着表面缩小。Ferenbaugh和Hindawi对菜豆叶片受酸雨胁迫的研究结果也是如此。由此可见，酸雨胁迫导致植物叶细胞受损、水分散失、叶片萎缩，大大降低了植物的抗性，

使害虫更容易入侵和取食。

4 展望

（1）酸雨对植物—害虫—天敌系统影响的研究，以往大多是在模拟酸雨和实验室条件下进行的，它对自然酸雨作用下植物、昆虫、天敌的个体发育、营养成分、形态结构、行为、繁殖以及种群的增长与衰退、群落结构的变化演替等很难做出准确的判断。其原因一是已有的实验大多数是酸雨的短期效应；二是自然降雨并非固定同一酸度，酸雨频次不一，酸雨与常规雨水（pH>5.6）交替进行；三是植物—害虫—天敌系统受酸雨在内的多种污染物和环境因子的综合作用，仅仅考虑酸雨一个因素的改变不能反映多种因子的联合作用。正是因为以上原因，今后的研究应该把重点转向野外和田间的长期监测。

（2）酸雨对植物—害虫—天敌系统影响的研究，相对于酸雨对植物—土壤系统和其他空气污染对植物—昆虫系统的研究来说起步较晚，研究较少。已有的研究仅仅从酸雨对害虫生长发育、取食量、消化率以及对植物营养物质的影响等方面做了部分工作。酸雨对昆虫的直接作用，酸雨对植物的挥发性物质、防御性物质、形态结构及其对昆虫生理生化的影响研究尚不深入，尤其是昆虫受酸雨胁迫后，其行为、新陈代谢、抗性及生态适合度的变化鲜有报道。这表明该领域还有很广阔的研究空间，它的理论意义和实践价值必将随研究的发展而逐渐显示出来。

（3）植物—害虫—天敌系统是一个复杂的系统，不仅要从酸雨作用于系统后植物营养物质、形态结构的变化，害虫与天敌的生理、行为变化、种群动态的变化以及它们之间的相互关系方面进行研究，而且还应从分子遗传水平上去研究酸雨胁迫后，昆虫行为和代谢的变化是短期可逆的应激反应还是可遗传的变异，探索酸雨对植物—害虫—天敌系统的作用机理，丰富环境因子对昆虫胁迫机理的研究内容。

原文刊载于《昆虫知识》，2004年第1期

温度对竹盲走螨实验种群生长发育与繁殖的影响

刘怀,赵志模,邓永学,徐学勤,李映平

摘要:在 16、20、24、28、32、36 ℃的 6 种恒温条件下,以竹裂爪螨各螨态为饲料,系统研究了温度对竹盲走螨实验种群生长发育、繁殖的影响。结果表明,36 ℃条件下竹盲走螨不能完成世代发育。在 16~32 ℃范围内,各螨态的发育历期随温度升高而缩短,卵至成螨的发育历期在 16、20、24、28、32 ℃ 5 种恒温条件下分别为 16.01、10.80、5.81、5.12 和 4.68 d;世代发育起点温度和有效积温分别为 10.44 ℃和 93.77 d·℃。分别用王-兰-丁模型和 Logistic 模型拟合了温度与发育速率的关系,其中王-兰-丁模型能更好地反映出竹盲走螨在高温下发育受到抑制的现象。产卵期和雌成螨寿命随温度的升高而缩短。平均每雌总产卵量在 24 ℃最高,为 31.26粒,32 ℃条件下仅为 9.20 粒,平均每雌总产卵量(Y)与温度(T)之间呈二次抛物线关系:$Y=-171.694\ 2+17.273\ 2T-0.036\ 3T^2$ ($R=0.964\ 8^*$)。用年龄等级法分析表明 16 ℃、32 ℃条件不利于竹盲走螨种群的生长发育和繁殖。

关键词:竹盲走螨;温度;生长发育;繁殖

Temperature Effect on Development and Reproduction of Experimental Population of *Typhlodromus bambusae* (Acari:Phytoseiidae)

Liu Huai, Zhao Zhimo, Deng Yongxue, Xu Xueqin, Li Yingping

Abstract: The Phytoseiidae Mite, *Typhlodromus bambusae*, was a dominant predator associated with the mite pests of moso bamboo. The effects of 6 constant temperatures (16, 20, 24, 28, 32 and 36 ℃) on development, oviposition and survival of *T. bambusae*, reared on *Schizotetranychus bambusae*, were studied. The eggs could not hatch at 36 ℃. Within the range of 16—32 ℃, the development time of egg, larva, protonymph, deutonymph shortened as temperature increased. The developmental threshold temperatures for egg, lava, protonymph, deutonymph, and from egg to adult were 10.21, 10.67, 10.94, 10.76, 10.44 ℃, respectively. The developmental effective accumulated temperatures for egg, lava, protonymph, deutonymph, and from egg to adult were 36.18, 12.61, 21.46, 21.63, 93.77 d℃, respectively. The nonlinear Wang-Lan-Ding model gave a good fit to the relationship between developmental rate and temperature. Percentage of survival was the highest (92.50%) at 20 ℃, and lowest (67.50%) at 32 ℃. Temperature affected fecundity and longevity significantly. The oviposition period was the longest (24.43 d) at 20 ℃, whereas the shortest (5.14 d) at 32 ℃. The eggs laid by per female ranged from 31.26 at 24 ℃ to 9.20 at 32 ℃. A nonlinear regression model could fit the relationship between eggs laid by per female (Y) and temperature (T). The equation was $Y=-171.694\ 2+17.273\ 2T-0.036\ 3T^2$ ($R=0.964\ 8^*$).

Key words: *Typhlodromus bambusae*; experimental population; development and reproduction; temperature

毛竹(*Phyllostachys heterocycla* cv. Pubescens)是我国南方最主要的经济树种,被林业部门定为退耕还林、保护环境、促进经济的主要森林树种。近年来,我国主要竹产区,如福建、浙江、四川等省的毛竹林出现大面积的竹叶黄化、落叶,甚至成片枯死的现象,各地区纷纷开展研究,从竹林老化、土壤衰退、大气污染、病虫发生等多个方面探讨枯死成灾的原因,最终认为是螨害,尤其是竹裂爪螨(*Schizotetranychus bambusae*)、食竹裂爪螨(*S. celarius*)等害螨类的危害是造成毛竹大面积枯死的主要因素。笔者于1998—2001年在四川省国家级风景名胜区"蜀南竹海"开展了毛竹害螨发生与防治的试验研究,并在林间实际调查发现竹盲走螨(*Typhlodromus bambusae*)是毛竹林重要的捕食性天敌,对毛竹的一些害螨有较强的控制作用。但有关竹盲走螨生物学、生态学以及对害螨控制等方面研究的文献资料较少,国外一些学者对其生物学方面进行了初步研究。在国内该螨为1996年报道的一个新纪录种,对其研究仅见张艳璇等的报道。本试验在前人研究的基础上,以毛竹害螨优势种竹裂爪螨为食料,系统探讨温度对竹盲走螨生长、发育、繁殖的影响,以期为利用该螨进行毛竹害螨的生物防治奠定基础。

1 材料与方法

1.1 供试虫源

在重庆北碚区缙云山毛竹林采集竹盲走螨,在室温条件下(温度25 ℃±0.5 ℃,

RH 70%～80%，L∶D=12∶12)用竹裂爪螨各螨态作为食料饲养若干代后作为供试螨源。竹裂爪螨的饲养采用植物叶片作为支持面的饲养方法，在室内常温下进行群集饲养，定期采集无病虫为害且颜色较绿的毛竹叶片作为饲料。

1.2 试验处理

试验共设置 16、20、24、28、32 及 36 ℃ 6 种温度，以 LRH-250-G 型光照培养箱(广东医疗器械厂)控制温度，温差变幅均为 ±0.5 ℃，RH 维持在70%～80%，每天给予 12 h 光照，采用植物叶片作为支持面的饲养方法进行饲养。在直径为 15 cm 的培养皿内铺吸满水的塑料泡沫，其上铺一层吸水滤纸，取长 13 cm 左右、宽 1 cm 左右的毛竹叶片平铺于滤纸上，叶背朝上，并且用脱脂棉条压住叶片的叶尖、叶柄以及围住叶片四周，然后移入竹裂爪螨雌成螨若干，让其自然产卵 2～5 d，然后移去部分雌成螨，保留 3～5头，使叶片上包含有竹裂爪螨卵、幼螨、若螨以及成螨等各螨态，以此作为饲养竹盲走螨的饲养皿。试验时，从群集饲养的竹盲走螨中，挑选若干头雌成螨于饲养皿的竹叶上，让其产卵，并每隔 8 h 检查 1 次，将 8 h 内所产卵视为同一时刻的初产卵，并将卵挑出，移入新的饲养皿叶片上，每叶片只挑入 1 粒卵，移至某一设置温度组内。进入试验时每温度组内的卵量为 40～50 粒。从移入卵时起，每隔 8 h 观察记录各处理的竹盲走螨的虫态变化、存活和死亡数、产卵量以及成螨寿命等。并根据各不同温度和螨态的情况，于每次检查时补若干竹裂爪螨的幼螨、若螨，保证每叶片上的食物包含有竹裂爪螨各螨态(主要以卵、幼螨和若螨为主)30～40 头(粒)，并根据不同温度每隔 3～5 d 更换 1 次含有竹裂爪螨新鲜竹叶。观察记录直到成螨死亡为止。

1.3 数据处理

试验所得数据在计算机上用"SPSS"软件进行统计分析、模型拟合。方差分析(F 测验)项目包括温度对竹盲走螨成虫寿命、产卵期、产卵量及存活率的影响，采用 Duncan 新复极差法进行多重比较。

2 结果与分析

2.1 不同温度条件下竹盲走螨各螨态发育历期

竹盲走螨个体发育经过卵、幼螨、前若螨、后若螨和成螨5个阶段,在蜕皮前无明显的静息期。雄螨比雌螨发育稍快,先发育成熟的雄螨常滞留于雌螨左右,待雌成螨成熟后与之交配,交配型为"钝绥螨—盲走螨类型"。不同温度下竹盲走螨的发育历期见表1。

表1说明,在16、20、24、28、32 ℃条件下,竹盲走螨卵至成螨历期分别为16.01、10.80、5.81、5.12和4.68 d。试验结果还表明,在36 ℃恒温条件下,竹盲走螨不能完成世代发育。在16~32 ℃范围内,卵期、幼若期以及世代历期都随着温度的升高而缩短,其中卵历期最长(5.97~1.92 d)、幼螨历期最短(2.17~0.62 d)。

发育起点温度和有效积温是昆虫种群的基本生物学参数之一。本研究采用直接最优法计算各螨态的发育起点温度及有效积温,结果见表2。从该表可以看出,竹盲走螨各螨态的发育起点温度约在10~11 ℃之间,其中卵至成螨的发育起点温度和有效积温分别为10.44 ℃和93.77 d℃。

表1 不同温度条件下竹盲走螨的发育历期

温度/℃	卵期/d	幼螨期/d	前若螨期/d	后若螨期/d	卵至成螨/d
16	5.97±0.61	2.17±0.22	3.86±0.40	3.98±0.32	16.01±0.73
20	4.08±0.48	1.56±0.21	2.87±0.23	2.57±0.33	10.80±0.69
24	2.20±0.25	0.82±0.21	1.38±0.27	1.46±0.24	5.81±0.39
28	1.92±0.22	0.71±0.14	1.24±0.15	1.24±0.19	5.12±0.29
32	1.92±0.27	0.62±0.14	1.08±0.20	1.08±0.22	4.68±0.39

表2 竹盲走螨各发育阶段的发育起点温度和有效积温

类型	发育阶段				
	卵期	幼螨期	前若螨期	后若螨期	卵至成螨
发育起点温度/℃	10.212 6	10.672 4	10.938 3	10.757 0	10.439 6
有效积温/(d℃)	36.184 5	12.611 6	21.464 5	21.628 5	93.765 5

2.2 各发育阶段的发育速率与温度关系的模拟

在昆虫(螨类)种群数量动态模拟中,一个很重要的模型是昆虫种群在各种温度条件下的发育率模型。根据表1,将发育历期转换成发育速率(V)后,用Logistic模型(公式1)及王-兰-丁模型(公式2)拟合竹盲走螨卵、幼螨、前若螨、后若螨、卵至成螨以及产卵前期等各发育阶段的发育速率(V)与温度(T)之间的关系,求解出的模型参数见表3,同时列出了各模型拟合效果的方差分析结果(用R^2表示)。两种模型的公式分别为:

$$V_1 = \frac{K}{1+\exp(a-bT)} \cdots\cdots\cdots\cdots\cdots\cdots\cdots\cdots\cdots\cdots (1)$$

$$V_2 = \frac{K}{1+\exp[-r(T-T_0)]}\left[1-\exp\left(\frac{T-T_L}{\delta}\right)\right]\left[1-\exp\left(\frac{T_H-T}{\delta}\right)\right] \cdots\cdots (2)$$

式(1)中,K:发育速率V的上限;T为温度;a、b为常数。式(2)中,K为潜在的饱和发育速率,等于最适发育速率$V(T_0)$的二倍;r是发育速率随温度变化的指数增长率;T_L、T_H为最低和最高临界温度;T_0为最适发育温度;δ为边界层宽度,其相对大小反映生物对极端温度的忍耐程度。

由表3可见,竹盲走螨卵、幼螨、前若螨、后若螨以及卵至成螨的发育速率与温度关系的Logistic模型和王-兰-丁模型拟合效果基本一致;但王-兰-丁模型能更好地反映出竹盲走螨在高温下发育受到抑制的现象,其检验模型拟合的相关指数(R)均大于Logistic模型的相关指数。并由王-兰-丁模型推知,竹盲走螨后若螨期(未成熟阶段)的发育最低、最高临界温度分别为9.89 ℃与35.53 ℃,最适发育温度为20.96 ℃(表3),这与本试验在36 ℃恒温条件下竹盲走螨不能完成世代发育的结果一致;各发育阶段的发育最低临界温度与有效积温法则计算出的发育起点温度也基本一致。

2.3 温度对竹盲走螨存活的影响

试验表明,竹盲走螨各发育阶段的存活率与温度有着密切的关系。从表4可以看出,在试验温度范围内(16～32 ℃),竹盲走螨各发育阶段的存活率均较高。温度对竹盲走螨幼螨的存活率几乎没有影响,其存活率均大于94.00%,20 ℃时最高,达100.00%,其原因可能是幼螨期发育历期短,对食物的要求较低，或者根本就不取食，可直接脱皮变成前若螨。从温度对竹盲走螨各发育阶段存活率的影响还可以看出，卵期和前若螨期的竹盲走螨对温度的变化表现得较为敏感。卵的孵化率在20 ℃和28 ℃条件下最高，均为97.50%，32 ℃条件下为85.00%；前若螨期存活率20 ℃时最高，为97.44%，16 ℃时较低，为82.86%。16、20、24、28、32 ℃各温度条件下卵至成螨的存活率分别为70.00%、92.50%、87.50%、85.00%和67.50%。

表3 竹盲走螨发育速率与温度关系的模型参数估计

模型		发育阶段					
		卵期	幼螨期	前若螨期	后若螨期	卵至成螨	产卵前期
王-兰-丁模型	K	0.816 4	2.190 0	1.263 0	0.233 5	1.322 0	0.645 7
	r	0.169 1	0.167 7	0.170 3	0.238 4	0.154 0	0.379 1
	T_0/℃	23.83	24.23	24.05	20.96	24.25	20.11
	T_L/℃	11.49	10.00	10.12	9.89	10.27	10.18
	T_H/℃	34.85	35.45	36.49	35.53	38.55	33.86
	δ	1.943 0	1.535 0	2.081	0.670 1	3.092 1	4.565 0
	R^2	0.985 9	0.971 7	0.980 5	0.986 7	0.989 4	0.992 7
Logistic模型	k	0.554 8	1.742 2	1.002 9	0.222 4	1.007 3	0.371 0
	a	5.690 4	4.826 7	4.957 6	5.204 0	4.842 7	12.475 0
	b	0.287 8	0.226 0	0.231 9	0.254 3	0.226 9	0.683 4
	R^2	0.965 9	0.968 9	0.974 1	0.976 0	0.988 9	0.622 4

表4 不同温度下竹盲走螨各发育阶段的存活率

发育阶段	温度/℃				
	16	20	24	28	32
卵期/%	92.50	97.50	95.00	97.50	85.00
幼螨期/%	94.59	100.00	97.36	94.87	94.12
前若螨期/%	82.86	97.44	97.30	94.59	93.73

续表

发育阶段	温度/℃				
	16	20	24	28	32
后若螨期/%	96.55	97.37	97.22	97.14	90.00
卵至成螨/%	70.00	92.50	87.50	85.00	67.50
产卵前期/%	81.13	95.45	100.00	90.48	88.24

2.4 温度对成螨寿命及繁殖力的影响

不同温度条件下竹盲走螨雌成螨的产卵前期、产卵期、产卵后期、寿命以及平均每雌总产卵量和平均每雌每日产卵量见表5。

从表5的结果可以看出,在70%~80%的相对湿度条件下,竹盲走螨雌成螨寿命在16℃条件下最长,平均为34.64 d,随着温度的升高,寿命逐渐缩短,32℃条件下最短,仅为10.76 d。在16℃低温条件下成螨的产卵前期和产卵后期较长,分别为11.86 d和5.46 d,是该温度下雌成螨寿命延长的主要原因。且适宜的温度范围内,产卵前期随温度的升高而逐渐缩短,而超过某一温度后,其发育历期反而延长;在16~32℃范围内产卵后期随温度升高而缩短。

竹盲走螨后若螨蜕皮成为成螨后即可进行交配。在试验温度范围内,经过2.16~11.86 d后雌成螨开始产卵,由表5可以看出,竹盲走螨平均每雌总产卵量在20℃与24℃2种温度条件下较高,分别为30.86粒和31.26粒,其次28℃时为28.35粒,但3种温度下产卵量的差异不显著。32℃条件下平均每雌总产卵量最低,仅为9.20粒。平均每雌每日产卵量表现为28℃最高,为2.33粒,而20℃与16℃分别为1.28和0.63粒,32℃下反而较高,为1.74粒,但32℃条件下产卵期最短,仅为5.14 d,是导致平均每雌总产卵量最低的直接原因。温度对竹盲走螨平均每雌总产卵量和平均每雌每日产卵量影响的变化趋势都呈开口向下的抛物线形式(如图1、图2)。因此,可用二次抛物线方程拟合二者与温度(T)的关系,其方程为:

平均每雌总产卵量(Y):$Y=-171.694\ 2+17.273\ 2T-0.036\ 31T^2$($R=0.964\ 8^*$,$F_{(2,4)}=36.379\ 6$,$P=0.026$);

平均每雌每日产卵量(Y): $Y=-8.4013+0.7933T-0.0148T^2$ ($R=0.9706^*$, $F_{(2,4)}=20.1356$, $P=0.047$)。

表5 温度对竹盲走螨寿命及繁殖力的影响①

项目	温度/℃				
	16	20	24	28	32
寿命/d	34.64±5.60Aa	31.54±7.60Aa	17.93±5.38Bb	15.61±4.13BCb	10.76±2.97Cc
产卵前期/d	11.86±1.84Aa	3.92±1.12Bb	2.16±0.45Cc	2.67±4.5Cc	4.49±1.77Bb
产卵期/d	17.38±5.01Bb	24.43±6.33Aa	14.37±5.20BCbc	12.10±4.20Cc	5.14±2.28Dd
产卵后期/d	5.46±1.80Aa	3.67±1.68Bb	1.42±0.51Cc	1.18±0.67Cc	1.13±0.69Cc
平均每雌总产卵量/粒	11±5.10Bb	30.86±7.72Aa	31.26±9.32Aa	28.35±10.40Aa	9.20±5.34Bb
平均每雌每日产卵量/粒	0.63±0.16Dd	1.28±0.11Cc	2.23±0.24Aa	2.33±0.17Aa	1.74±0.27Bb

①表中同一行标有不同字母分别表示不同温度处理间差异表达极显著($P<0.01$)或显著($P<0.05$)水平。

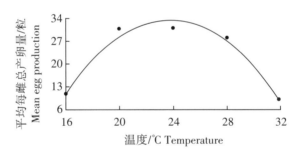

图1 竹盲走螨平均每雌总产卵量与温度的关系

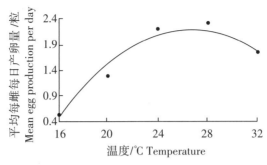

图2 竹盲走螨平均每雌每日产卵量与温度的关系

2.5 不同温度下竹盲走螨的产卵特性

由于不同恒温条件下的产卵期长短差异较大，平均每雌每日产卵量不能准确反映出温度变化对繁殖力所产生的影响。为了在同一水平上比较其繁殖力与温度的关系，现将不同温度下的产卵期划分成10个年龄级，具体划分情况如下，16 ℃条件下每3 d作为1个等级（产卵期最长为32 d），20 ℃每4 d作为1个等级（产卵期最长为41 d），24 ℃每3 d作为1个等级（产卵期最长为33 d），28 ℃每2 d作为1个等级（产卵期最长为19 d），32 ℃每1 d作为一个等级（产卵期最长为12 d），用每一等级内平均每雌日产卵量（Y）作为指标来衡量其繁殖力与温度的关系（见图3），其中：$Y=N/(n \cdot d)$。式中N为每一等级内的总产卵量，n为雌螨数，d为等级间隔天数。

由图3可以看出，当温度较低时（16 ℃），日产卵量少，生殖高峰不明显。20～28 ℃时，产卵由少到多，在第2年龄等级就可形成生殖高峰，其中24 ℃、28 ℃条件下产卵高峰期持续时间长，到第5年龄等级日平均产卵量才逐渐下降。32 ℃条件下，产卵期短，生殖高峰明显后移。

3 讨论

在供试的温度（16～36 ℃）范围内，36 ℃条件下，竹盲走螨种群不能够完成世代发育，主要是因为卵在36 ℃的恒温条件下不能孵化。本试验将24 ℃条件下产生的初孵幼螨移入36 ℃恒温下进行饲养，其结果表现为各发育阶段存活率低，发育历期延长，即使有少量个体能发育至成螨，但都不能正常繁殖产卵,这与植绥螨的一些种类对温度的要求基本一致,如吴伟南等研究发现当温度超过32 ℃时,智利小植绥螨生长发育将受到抑制。

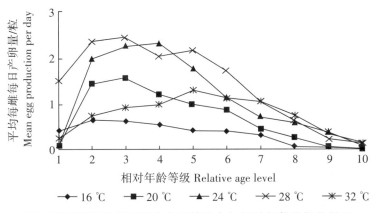

图3 不同温度条件下竹盲走螨繁殖力与相对年龄等级的关系

试验研究表明在16～32 ℃之间,随着温度升高,竹盲走螨的卵、幼螨、若螨的发育历期都逐渐缩短,亦即发育速率逐渐加快,但产卵前期在32 ℃延长。Saito对竹盲走螨的生物学进行过初步研究,结果表明在(25±0.5)℃的恒温条件下,以食竹裂爪螨为食时,卵到成螨的发育历期雄螨为6.1 d,雌螨为7.3 d;而张艳璇等研究表明,在22～30 ℃的室温(变温)条件下以南京裂爪螨为食时,卵期为1.7 d,幼螨期为1.0 d,前若螨期为0.8 d,后若螨期0.8 d;本研究以竹裂爪螨为食时,(24±0.5)℃的恒温条件下,卵到成螨的发育历期5.8 d,形成差异可能是不同的地理种群以及不同食料营养差异所致。

昆虫(螨类)发育速率与温度的关系是昆虫或螨类生物学研究的主要内容,它对于害虫害螨预测预报和益虫的利用有着巨大的价值,历来受到人们的重视。本研究分别用生态学中的基本模型Logistic方程和王-兰-丁模型拟合了竹盲走螨生长发育速率与温度的关系,其结果表明王-兰-丁模型能更好地反映不同温度条件下竹盲走螨生长发育情况,根据该模型推导出的各发育阶段对温度临界值与最适值和试验观察结果基本一致。

种群的死亡率(或存活率)、繁殖力、平均寿命是影响种群数量动态的基本生物学特征。温度作为一个重要的生态因子,作用于种群系统,必然引起上述基本生物学特征发生变化。本试验在16~36 ℃的恒温条件下研究了温度对竹盲走螨存活、繁殖力以及平均寿命的影响。结果表明,竹盲走螨卵至成螨的存活率以20 ℃最高,达92.50%。各虫态比较而言,卵和前若螨期存活率对温度的变化表现得较为敏感,高温不利于卵的孵化,其中卵期的存活率在32 ℃时仅为85%,而其他温度处理均在92%以上。雌成螨的平均寿命在16 ℃最长,为34.64 d,随温度的升高,其寿命缩短。方差分析表明,16 ℃与20 ℃条件下的雌成螨寿命(分别为34.64和31.54 d),差异不明显。平均每雌总产卵量在20 ℃与24 ℃较高,分别为30.86粒与31.26粒。本文利用年龄等级法,以每一等级内平均每雌每日产卵量作为指标反映了种群增长和繁殖力特征与温度的关系,结果表明温度较低时(本试验在16 ℃),虽然雌成螨寿命与产卵期较长,但日产卵量少,不能产生明显的生殖高峰,32 ℃的较高温条件下,虽日产量较高,但产卵期短,日生殖高峰明显后移,即低温和高温均表现出不利于种群的增长。

原文刊载于《林业科学》,2004年第1期

论文选集

主题二

农产品储运保护研究

双低储粮虫螨群落组成研究

邓永学,吴仕源,赵志模

摘要:1990—1992年对四川合川和广安县小麦双低仓虫螨群落组成调查表明,腐食酪螨及书虱为粮仓虫螨群落优势种群。模拟双低储藏与实仓调查结果一致。

关键词:昆虫;螨目/双低储藏;储粮虫螨;群落

A Preliminary Study on the Community Composition of Insects and Mites of Stored Wheat in Low Oxygen and Low Pesticide Environment

Deng Yongxue, Wu Shiyuan, Zhao Zhimo

Abstract: Investigations of the community composition in wheat storehouses of low oxygen and low pesticide environment made in 1990—1992 in Hechuan city and Guang'an county of Sichuan province showed that *Tyrophagus putrescentiae* and *Liposcelis bostrychophila* constituted the dominant population. The results of a simulated experiment with low oxygen and low pesticide environment and an on-the-spot investigation were consistent with each other.

Key words: insects; Acarina/stored in low oxygen and low pesticide; insects and mites of stored wheat; community composition

近年来,在双低储粮中,害虫害螨抗性日渐突出。本研究对小麦双低仓虫螨群落组成进行了调查,并采用PVC塑料薄膜密闭。模拟双低储藏,观察虫螨群落消长动态。

1 材料和方法

1.1 实仓调查

采用5点取样法,调查四川省广安县白塔粮仓及合川市白鹿小麦粮库内上、中、下层小麦各1 kg,观察计数害虫害螨的种类、数量和粮仓温度、粮食水分、气体浓度等,每月1次,周期1年。

1.2 PVC模拟仓储调查

采用PVC塑料薄膜作密闭材料,设自然密闭降O_2,密闭降O_2+磷化铝熏蒸1.5 g/m³,马拉硫磷20 mg/kg+密闭降O_2,马拉硫磷75 mg/kg+密闭降O_2及纸箱储藏5种处理,每处理装小麦50 kg,重复3次。调查项目同上。

2 结果和分析

2.1 仓库害虫群落组成

调查表明,两地小麦仓库害虫害螨群落组成简单,优势种群为腐食酪螨 *Tyrophagus putrescentiae*、害嗜鳞螨 *Lepidoglyphus destructor* 及书虱 *Liposcelis bostrychophila*,3种害虫在粮仓密闭降O_2及施药后均可存活繁殖,且数量较大;粮仓四角数量多于中央,上层多于中层,下层数量极少。年消长与粮仓温度,粮食水分,O_2及CO_2浓度,施药因素等有关。在小麦双低仓虫螨群落中,次要种群有长角扁谷盗 *Cryptolestes pusillus*、玉米象 *Sitophilus zeamais*、日本蛛甲 *Ptinus japonicus*、拟裸蛛甲 *Gibbium aequinoctiale*、锯谷盗 *Oryzaephilus surinamensis*、杂拟谷盗 *Tribolium confusum* 等。其中,玉米象及日本蛛甲在周年中均发现各1头,长角扁谷盗及锯谷盗种群数量较大。未发现鳞翅目害虫及谷蠹 *Rhyzopertha dominica*。白塔粮库较白鹿粮库虫螨种类多,数量大,这与两地粮食保管条件有关。

2.2 模拟储藏

采用马拉硫磷20 mg/kg,75 mg/kg处理和塑料薄膜密封储藏,能有效地防治玉米象和谷蠹。1个月后,O_2浓度可降低14.6%,CO_2浓度可升至5.8%,但对腐食酪螨及书虱的防治效果不佳,次年大量繁殖,种群迅速增大,成为优势种群。

按1.5 g/m³投磷化铝密闭储藏的,其虫螨群落组成与马拉硫磷处理结果一致，但粮堆中 CO_2 浓度较低，O_2 浓度较高，这可能是 PH_3 气体杀死了粮堆中具呼吸作用的各种有机体所致。

仅用 PVC 塑料薄膜密闭，粮堆中各种有机体呼吸作用旺盛，CO_2 浓度迅速上升，O_2 浓度迅速下降，主要储粮害虫玉米象、谷蠹、麦蛾 Sitotroga cerealella、印度谷螟 Plodia interpunctella 因缺 O_2 而死亡，但仍可发现腐食酪螨及书虱。储藏超过半年，由于粮堆密闭效果逐渐变差，O_2 浓度回升，腐食酪螨及书虱种群数量迅速增长，成为优势种群。笔者认为，此种方法不用药，经济安全，能有效地控制主要储粮害虫，因此值得进一步研究推广。

采用低箱储藏，害虫害螨及天敌种类多，数量大，粮食受害严重。其虫螨群落组成有：麦蛾、印度谷螟、粉斑螟 Ephestia cautella、四点谷蛾 Tinea tugurialis、玉米象、谷蠹、杂拟谷盗、锯谷盗、长角扁谷盗、小菌虫 Alphitobius Laevigatus、书虱、害嗜鳞螨、腐食酪螨、圆腹宽缝拟蝎 Chelifer panzeri、米象金小蜂 Lariophagus distinguendus 及捕食螨等。其群落演替具一定规律，即由初级仓虫至次级仓虫再至 3 级仓虫。

3 讨论

在双低储粮中，低氧、低药对储藏物害虫及微生物的活动起着抑制作用。低氧能控制储粮害虫和好气性微生物的呼吸作用，恶化其生态条件。低药熏蒸，刺激微弱，害虫气门不会立即关闭，氧气和 PH_3 随之进入虫体，从而有效地抑制线粒体的呼吸作用，破坏细胞的生化功能，使昆虫的代谢作用停滞而中毒死亡。在双低储粮中，腐食酪螨及书虱抗药性日渐突出，应加强其抗性机制研究，以确保粮食安全储藏。

原文刊载于《西南农业大学学报》，1995 年第 1 期

腐食酪螨对低氧高二氧化碳气调的抗性

赵志模,张肖薇

摘要:在 10% CO_2,5% O_2(其余部分为 N_2,下同);35% CO_2,11% O_2;35% CO_2,21% O_2;75% CO_2,11% O_2 及 75% CO_2,21% O_2 五种气调环境下对腐食酪螨 *Tyrophagus putrescentiae*(Schrank)进行连续 25~26 代的抗气性诱导筛选。结果表明,抗性都有不同程度的增加,在 75% CO_2,11% O_2 下抗性系数最大为 5.0,在 10% CO_2,5% O_2 下抗性系数最小为 3.7,在其他气调环境下,抗性系数都在 4.0 以上。

关键词:腐食酪螨;气调;抗性诱导

Development of Resistance of *Tyrophagus putrescentiae* to Controlled Atmosphere

Zhao Zhimo, Zhang Xiaowei

Abstract: Laboratory induced resistance of successive 25~26 generations of *Tyrophagus putrescentiae* to the gas under the controlled atmosphere (N_2 as the balance gas) as follows; 10% CO_2, 5% O_2; 35% CO_2, 11% O_2; 35% CO_2, 21% O_2; 75% CO_2, 11% O_2; 75% CO_2, 21% O_2. The results showed the resistant ability to the gas has been increased. RF (Resistant Factor) was the highest 5.0 under 75% CO_2, 11% O_2; The lowest 3.7 under 10% CO_2, 5% O_2; the other RFs were up to 4.0. All increasesments of resistance can be expressed by Logistical equation.

Key Words: *Tyrophagus putrescentiae*; controlled atmosphere; development of resistance

仓虫(螨)对熏蒸剂等化学药剂的抗性早为人们所注意,并已经在这方面做了大量的研究工作。气调储粮技术对于仓虫(螨)的防治有明显的效果,但近年来,仓虫(螨)对低 O_2 高 CO_2 气调的抗性问题引起了国内外仓虫专家的关注,如 Vonahaye 1990 年报道了在实验室条件下对赤拟谷盗抗性的诱导情况,Bond,E.J. 1979 年诱导了谷象对 CO_2 的抗性。本实验是在几种气调环境下,诱导腐食酪螨的抗气性能力,考察气体浓度和抗性增长趋势间的关系,为气调储粮和仓螨防治提供依据。

材料和方法

供试虫源来自西南农大仓虫生态研究室用面粉和酵母粉(3∶1)混合饲养多代的腐食酪螨纯系。

气调设备：用钢瓶贮气(气体浓度 $N_2>99.9\%$，$O_2>99.2\%$，$CO_2>99.0\%$)，气体经减压阀，稳流阀，流量调节阀，转子流量计进入混气瓶，然后由乳胶管导入 1 000 mL 的密闭广口瓶内，每次充气 15 min。按实验要求以 N_2 作平衡气体调节 CO_2 和 O_2 的比例。

(1)标准 LT_{50} 的测定

为了比较各处理螨的抗性增长情况，本实验对诱导过程中螨的 LT_{50} 都采用统一的测定方法，用该法测定的 LT_{50} 都称标准 LT_{50}，其测定方法是：将气体含量为 55% CO_2，11% O_2 的密闭广口瓶置于 28 ℃，75%～85% RH 下，每瓶装入高 1 cm、直径 2 cm 的 3 个小塑料盒(盒盖由 160 目尼龙筛网做成)，每盒接 10 头成螨，共 6 个广口瓶，当螨的死亡率达 10% 时开始，以后每隔一段时间开启一瓶作记录，最后一次死亡率约 90% 左右。重复 3 次，取死亡率在 10%～90% 的时间值，采用最小二乘法求 LT_{50}。

(2)气调的抗性诱导

诱导前先测定供试螨种群的标准 LT_{50}，然后在 28 ℃，75%～85% RH 下，设置五组处理即 10% CO_2，5% O_2；35% CO_2，11% O_2；35% CO_2，21% O_2；75% CO_2，11% O_2；75% CO_2，21% O_2。各处理每周充气一次。第一组处理不打开瓶口敞气，第二、三组处理每次充气后的 8 h 敞气，第四、五组处理由于 CO_2 浓度较高，于充气后 4 h 敞气。各处理设 2 个重复，总样本量约 4 000 头。每隔 6 周测定一次各处理的标准 LT_{50}。

结果与分析

诱导前没有抗气性的品系的标准 LT_{50} 为 23.8 h。在 28 ℃，75%～85% RH 下，该螨完成一个世代大约需要两周，本实验共诱导 25～26 代，各处理测定九次标准 LT_{50}，每次测定间隔约 2～3 代。各次测定结果和各气调环境下最终的抗性系数(Resistant Factor)见表 1。

从表中可见，在75% CO_2，11% O_2下，诱导25～26代后，该螨的标准LT_{50}增加为原来的5倍，高于其他气调环境下诱导的RF；其次是35% CO_2，11% O_2和75% CO_2，21% O_2，RF分别为4.7和4.6；再次是35% CO_2，21% O_2，RF为4.1；在10% CO_2，5% O_2下RF最低，为3.7。在相同CO_2浓度下，O_2浓度较低的组，RF值较大；在相同O_2浓度下，CO_2浓度较高的组，RF值较大。第五组的O_2浓度与自然大气的相近为21%，这组的选择压力主要是来自75%的高浓度CO_2，该组的螨没有受到低O_2的选择压力。第二组的CO_2不及第五组的高，且O_2浓度仅是第五组的一半，该组的螨同时受到低O_2低CO_2的选择压力，因而第二组和第五组的RF值相近。

表1　5种气调环境下螨的RF值和各次测定的标准LT_{50}

测项	处理组别	一	二	三	四	五
气体成分/%	CO_2	10	35	35	75	75
	O_2	5	11	21	11	21
各次测定标准LT_{50}/h	1	32.6	35.4	37.5	41.7	39.7
	2	41.2	56.7	45.9	60.4	53.4
	3	49.6	71.1	56.5	71.0	2.9
	4	58.9	78.6	64.5	82.0	80.4
	5	66.4	88.7	72.9	92.9	88.5
	6	72.5	92.7	82.5	97.7	92.5
	7	80.8	96.4	86.3	104.6	96.8
	8	85.6	104.4	93.1	112.0	106.0
	9	86.9	111.3	96.5	119.0	110.1
RF		3.7	4.7	4.1	5.0	4.6

注：$RF = \dfrac{诱导终了的LT_{50}}{诱导前的LT_{50}}$；诱导前标准的$LT_{50}$为23.8 h。

从表1还可看出,在诱导开始的最初几代,LT_{50}增加的幅度较大,以后随着代数的增加,LT_{50}增加的幅度愈来愈小。这种情况的出现可能是因为螨在刚进入诱导阶段所受的选择压力要大些,不适者被大量淘汰,种群的抗性明显提高,随着诱导时间的延长,螨逐渐适应气调环境,选择压力相对减弱,淘汰的个体数减少,抗性提高的速率减慢。抗气性的这种增长趋势可以用逻辑斯蒂方程卡方(χ^2)检验如下:

处理1:$LT_{50} = \dfrac{100.04}{1+\exp(0.674-0.281T)}$ $\chi^2 = 2.89$

处理2:$LT_{50} = \dfrac{119.84}{1+\exp(0.628-0.322T)}$ $\chi^2 = 5.19$

处理3:$LT_{50} = \dfrac{107.11}{1+\exp(0.643-0.301T)}$ $\chi^2 = 2.52$

处理4:$LT_{50} = \dfrac{128.62}{1+\exp(0.635-0.317T)}$ $\chi^2 = 3.50$

处理5:$LT_{50} = \dfrac{112.15}{1+\exp(0.816-0.448T)}$ $\chi^2 = 2.68$

查表得$\chi^2(0.05,8)=15.51$,以上方程的$\chi^2<\chi^2(0.05,8)$,说明该螨对气调的抗性呈逻辑斯蒂增长。该方程表明,在一定气调环境下,抗性增长是有一定限度的,越接近最大限度值,抗性增长越缓慢。抗性增长的最大限度值(K),和选择压力有关,压力越大,K值越大。

原文刊载于《蛛形学报》,1993年第2期

磷化氢熏蒸处理对嗜卷书虱不同虫态的致死作用

丁伟,陶卉英,张永强,王进军,赵志模

摘要:实验室条件下系统研究了磷化氢(PH_3)对储物害虫嗜卷书虱 *Liposcelis bostrychophila* Badonnel 卵、各龄若虫和成虫的致死作用,并选用 PH_3 间歇熏蒸以及 PH_3 与气调交替处理等措施对嗜卷书虱进行处理,比较了不同处理措施对嗜卷书虱种群的控制效果。结果表明,PH_3 熏蒸处理对嗜卷书虱各虫态有不同的致死效果。对卵而言,24、72 和 120 h 熏蒸处理的 LC_{50} 分别为 0.137、0.045 和 0.035 mg/L;而 24 h 熏蒸处理对若虫的 LC_{50} 在 4.285~7.364 μg/L 之间,对成虫的 LC_{50} 为 20.404 μg/L;采用 25 μg/L 的 PH_3 进行 24 h 熏蒸处理,间隔 10 d 后再分别进行第 2 次和第 3 次熏蒸处理,可以完全控制嗜卷书虱的发生。采用 PH_3(12 μg/L)和气调(体积比例为 35% CO_2,1% O_2,64% N_2)交替处理能够延缓嗜卷书虱种群抗性的发展,交替处理 3~5 次可以完全控制嗜卷书虱的发生。

关键词:嗜卷书虱;磷化氢;气调处理;储物害虫

Lethal Effect of Phosphine on Different Stages of *Liposcelis bostrychophila* Badonnel (Psocoptera: Liposcelididae)

Ding Wei, Tao Huiying, Zhang Yongqiang, Wang Jinjun, Zhao Zhimo

Abstract: Lethal effects of phosphine (PH_3) on egg, nymph and adult of *L. bostrychophila* Badonnel were conducted in laboratory and lethal concentration was systematically determined in this research. In addition, *L. bostrychophila* was treated with PH_3 intermittent fumigation and alternative treatment of controlled atmosphere (CA) and PH_3. And then compare the control efficiency of different treatments on *L. bostrychophila* population. Results showed that PH_3 fumigation had different lethal effects on different stages of *L. bostrychophila*. After 24, 72 and 120 h exposure, LC_{50} of egg were 0.137, 0.045, and 0.035 mg/L, respectively; after 24 h treatment, LC_{50} of nymph varied in the rage of 4.285—7.364 μg/L, and that of adults was 20.404 μg/L. Exposed to 25 μg/L PH_3 for 24 h every 10 d, the population was completely controlled. And after alternative treatment with PH_3 (12 μg/L) and CA (35% CO_2, 1% O_2, 64% N_2), the resistance of population was delayed, and the occurrence of *L. bostrychophila* was completely controlled after the alternative treatment for 3—5 times.

Key words: *Liposcelis bostrychophila* Badonnel; PH_3; CA; stored product insect pest

磷化氢(PH_3)是一种广泛应用于储物害虫防治的熏蒸药剂,特别是用于储藏期较长的粮食和其他农产品。PH_3使用方便、效果良好、渗透力强,在粮仓中残留毒性低,加之目前还没有更好的备选熏蒸药剂,因而PH_3仍是储藏物害虫治理中不可或缺的一种重要药剂。但是,长期单一使用PH_3容易诱发害虫的抗性,并导致用药量的不断增加。因此研究PH_3对主要储物害虫熏蒸致死作用的特点,选择恰当的施药技术,是目前储物害虫化学控制的重要课题之一。

书虱(booklouse)是广泛存在于储藏物中的一类微小昆虫,大量发生时可造成严重危害,而且已发展成我国大型粮仓中的优势害虫种群。实践证明PH_3可以用来控制书虱的危害,Pike的研究证明嗜虫书虱 *Liposcelis entomophila* 对PH_3比较敏感,120 h时对卵的LD_{99}为1.66 mg/L。但是在实际应用中,磷化氢熏蒸处理后,书虱的种群数量很快就会得以恢复,这成为PH_3用于书虱治理的一个重要问题。关于PH_3熏蒸处理对嗜卷书虱的致死作用,特别是对不同虫态的影响,目前还没有相关的研究报道。本实验系统研究了PH_3对嗜卷书虱不同虫态的致死作用,并选用不同的处理措施对嗜卷书虱的种群进行控制,以便为进一步研究书虱对PH_3的忍耐性和抗性机理、寻找书虱抗性治理的有效措施提供依据。

1 材料与方法

1.1 供试虫源

嗜卷书虱 *Liposcelis bostrychophila* Badonnel,1990年采自西南农业大学应用昆虫及螨类生态研究室的模拟粮仓中,是在实验室27 ℃±0.5 ℃、相对湿度(RH)75%~80%、不予光照的条件下,以全麦粉、酵母粉、脱脂奶粉为原料配制的混合饲料饲养多代获得的品系。饲养方法参照Leong,Ho和丁伟等的方法。该品系在饲养过程中,培养条件稳定,没有接触任何化学农药,因此将该品系视为敏感品系的供试虫源。

1.2 供试药剂

56%的磷化铝（AlP）片剂（山东济宁化工厂生产）。磷化铝片剂在水分的作用下产生PH_3气体,将气体收集在密闭的集气瓶中,用硝酸银处理过的硅胶管检测含量后,用微量注射器移出所需的气体量供试。PH_3相对分子质量34,沸点$-87.5\ ℃$,30 ℃时蒸汽压为42.101×10^6 Pa,g/m^3换算为mL/m^3的换算系数由下式求得：$\frac{273+t}{273}\times660$,式中$t$为温度（℃）。施药时,采用微量注射器（规格50 μL或100 μL）,分别吸取一定体积的PH_3,注入测试瓶（1 000 mL广口瓶）中。

1.3 气调设备和条件选择

采用西南农业大学应用昆虫及螨类生态研究室设计组装的气体浓度控制仪。O_2、N_2、CO_2气体分别贮存于钢瓶中（气体纯度分别为$O_2>99.2\%$、$N_2>99.9\%$、$CO_2>99.0\%$）。参照吴仕源等的方法,以N_2作为平衡气体,按实验所需调节N_2、CO_2、O_2的流量,充气0.5 h后,使气体组配的体积分数为35% CO_2、21% O_2和44% N_2。所用容器为1 000 mL广口瓶。利用光照培养箱控制温度为27 ℃±0.5 ℃、RH为75% ~ 80%,气调处理过程中不予光照。

1.4 试验方法

1.4.1 杀卵作用测定

在直径2 cm的塑料养虫盒中放入少量饲料,挑入健康活泼的4龄若虫,每盒50头,于27 ℃±0.5 ℃、75% ~ 80% RH条件下培养。待若虫羽化后,将成虫转移到放有黑色滤纸的养虫盒中,产卵48 h后将成虫移走。将产有卵的养虫盒放入1 000 mL的熏蒸瓶中处理,于不同处理时间（24、72、120 h）分别打开熏蒸瓶,取出养虫盒,继续培养15 d后,根据各处理卵的孵化情况比较杀卵效果。

1.4.2 对若虫及成虫的致死作用测定

用细毛笔轻轻挑取一定量羽化后5 d的嗜卷书虱成虫或整齐一致的各龄若虫于直径为2 cm、高1 cm的塑料培养盒内,用160目纱网做成的盒盖盖严,放在铁丝编制的小铁笼中,置于1 000 mL广口瓶的中间部位。每个处理60头试虫,重复3次。然后分别通过瓶塞上的乳胶管注入不同体积的PH_3,再用止水夹将乳胶管夹住,密封后放置在27 ℃±0.5 ℃、RH 75%~80%的培养室内,24 h后移出培养盒,在27 ℃±0.5 ℃、RH 75%~80%的培养室内饲养10 d后,检查试虫死亡情况,计算各处理的校正死亡率。采用唐启义等的DPS处理系统,用机率值①分析法计算LC_{50}。

1.4.3 PH_3间歇熏蒸的处理方法

用与成虫致死作用测定相同的生测方法进行熏蒸处理。采用2龄若虫LC_{99}的PH_3浓度进行第1次熏蒸处理(24 h),24 h后打开熏蒸瓶,将试虫放置在27 ℃±0.5 ℃、RH 75%~80%的培养室内饲养,10 d后检查试虫的死亡情况,记录存活的虫口数量,然后重复进行第二和第三次熏蒸处理,每次熏蒸处理24 h。第三次熏蒸后15 d,检查各处理存活的虫口数量。

1.4.4 气调(CA)和PH_3熏蒸交替处理的试验设计与处理方法

参照Ding等的实验设计,将嗜卷书虱分成4个种群:

1)气调种群(CA-s)(35% CO_2,1% O_2,64% N_2):刚开始处理4 h,以后每次增加2 h;

2)PH_3种群(PH_3-s):以12 μg/L PH_3处理24 h而得到的种群;

3)气调和PH_3混合处理的种群(ALT-s):以气调(35% CO_2,1% O_2,64% N_2)和PH_3交替处理;

4)对照种群(CK-s):未经气调和药剂处理的种群。

以上每个种群处理5次,每14 d处理1次。每个处理至少重复10次,其

---

① 现机率多写作概率。——编辑注

中3次用以研究种群发展,其余的用来做生测实验。每次重复的试虫虫口基数为30头,每次处理后都将虫子置于27 ℃±0.5 ℃、RH 80%条件下饲养,24 h后检查活虫数。最后对处理次数和每次处理后的活虫数进行方差分析。

2 结果与分析

2.1 PH_3熏蒸处理对嗜卷书虱卵的致死作用

用PH_3对嗜卷书虱的卵进行熏蒸处理,根据处理剂量和时间不同,表现出不同的熏蒸杀伤作用。经过一段时间后,大部分卵干瘪死亡,但其余部分孵化出的若虫仍能正常存活。结果见表1。

表1 磷化氢对嗜卷书虱卵的校正死亡率(在干燥器中)*

单位:%

处理时间/h	磷化氢浓度/(mg·L^{-1})				
	0.02	0.03	0.06	0.12	0.25
24	12.2	26.3	30.6	47.8	62.3
72	21.5	49.2	62.1	71.2	82.5
120	29.3	44.5	73.2	86.4	91.6
192	100	100	100	100	100

注:*对照的平均死亡率为13%。

由表1可以看出,卵的死亡率随着处理时间和浓度的增加而增加。在0.25 mg/L的浓度下熏蒸处理24 h,卵的校正死亡率仅为62.3%,120 h也仅为91.6%;当处理时间达到192 h(8 d)时,即使在0.02 mg/L的条件下,也可以达到100%的致死效果。这可能是因为嗜卷书虱的卵一般在8 d左右即开始孵化,长时间的熏蒸处理,即使卵没有被杀死,卵孵化出的若虫也会被杀死,因此在很低的浓度下,也可以达到100%的致死效果。对表1中24、72、120 h杀卵效果的数据进行机率值分析,得表2。

表2 磷化氢处理嗜卷书虱卵的机率值分析(在干燥器中)[a]

处理时间/h	致死中浓度 LC$_{50}$			99%致死浓度 LC$_{99}$		
	磷化氢浓度/(mg·L^{-1})	χ^{2}[b] ($d.f.=3$)	FD[c]	磷化氢浓度/(mg·L^{-1})	χ^{2}[b] ($d.f.=3$)	FD[c]
24	0.137	2.605	0.156~0.118	10.798	2.605	17.699~3.876
72	0.045	8.858	0.050~0.040	2.270	8.858	3.249~1.290
120	0.035	3.744	0.038~0.032	0.587	3.744	0.733~0.442

注：a.暴露浓度范围0.02~0.25 mg·L^{-1}；b.所有χ^{2}值均表明没有显著差异；c.FD=95%置信区间，用于所述磷化氢浓度。

从表2可以看出，在120 h熏蒸处理时，达到99%的死亡率需要的浓度为0.587 mg/L；而24 h的熏蒸处理，要达到99%的死亡率，则需要的浓度为10.798 mg/L。虽然前者的熏蒸处理时间比后者增加了4倍，但处理所需的药剂浓度却减少了10.211 mg/L。

2.2 磷化氢熏蒸处理对成虫和若虫的急性致死作用

采用PH$_3$进行24 h熏蒸处理，测定PH$_3$对嗜卷书虱的致死作用，结果(见表3)表明：嗜卷书虱不同虫态对PH$_3$熏蒸处理的敏感性有很大的差异，以LC$_{50}$来评判，各虫态敏感性由高到低排列依次为1龄>2龄>4龄>3龄>成虫，其中成虫最不敏感，1龄若虫最为敏感；以LC$_{99}$来评判，各虫态敏感性由低到高排列依次为成虫>2龄>3龄>4龄>1龄，1龄和4龄若虫较为敏感，这可能与1龄若虫较小，4龄若虫的呼吸加强、代谢旺盛有关。对嗜卷书虱若虫来说，用LC$_{50}$ 95%的置信限来进行判断，则2龄和4龄若虫基本重合，而1龄若虫显著降低，3龄若虫显著增高。说明1龄、3龄若虫同2龄和4龄若虫之间对PH$_3$的敏感性存在着明显的差别。

表3　磷化氢对嗜卷书虱成虫和若虫的致死作用

龄期	回归直线 Y=	卡方值 χ^2	致死中浓度 LC_{50} 浓度 /($\mu g \cdot L^{-1}$)	95% FD	99%致死浓度 LC_{99} 浓度 /($\mu g \cdot L^{-1}$)	95% FD
1龄	2.440+4.052x	4.489	4.285±0.137	4.025~4.561	16.070±1.849	12.826~20.135
2龄	1.968+3.826x	9.194	6.202±0.191	5.839~6.589	25.152±3.421	19.267~32.836
3龄	0.831+4.807x	9.827	7.364±0.184	7.013~7.733	22.442±2.373	18.241~27.611
4龄	1.362+4.560x	16.022	6.278±0.166	5.960~6.613	20.323±2.099	16.598~24.883
成虫	−9.827+11.321x	1.213	20.404±0.224	19.970~20.849	32.751±0.955	30.931~34.678

2.3 磷化氢间歇熏蒸处理对嗜卷书虱的致死作用

采用 PH_3 对嗜卷书虱2龄若虫 LC_{99} 的浓度（25.0 μg/L），在实验室模拟实仓条件下对嗜卷书虱不同虫态的混合种群进行间歇熏蒸处理，共处理3次，结果见表4。

表4　磷化氢间歇处理对嗜卷书虱成虫和若虫的致死效果

龄期	第一次处理 处理样本数	死亡率/%	第二次处理 处理样本数	死亡率/%	第三次处理 处理样本数	死亡率/%
成虫	98	87.75	14	92.86	1	100.00
若虫	202	99.01	241	94.61	13	100.00

从表4中可以看出，采用高于所有虫态 LC_{50} 的 PH_3 浓度（25.0 μg/L），但低于最不敏感虫态成虫 LC_{99} 的 PH_3 浓度（32.8 μg/L）情况下，经过一次熏蒸处理，虽然总的虫口减退率为95.33%；但经过10 d后，虫口数量却达到255头，仅比处理前减少15%，这主要是由于采用25 μg/L的浓度进行熏蒸处理对卵没有明显的杀伤作用，第二次处理后，可以杀死大部分新孵化的幼虫和部分残存成虫，第三次处理后，可以完全将书虱的种群灭掉。

2.4 气调和PH_3交替处理对嗜卷书虱种群增长的影响

从表5可以看出,对照与其他几个处理相比较仍达到了显著水平($P<0.05$)。经CA和PH_3处理一次后,死亡率分别为16.7%和58.7%,但是存活的试虫仍能大量繁殖,经11周后,CA-s和PH_3-s数量分别增加了3.3倍和11.4倍。起初,PH_3-s的死亡率显著高于CA-s($F=51.62$;$df=3$,8;$P<0.0001$),经过11周后,PH_3-s的数量明显高于CA-s($F=633.41$;$df=3$,8;$P<0.0001$)。而采用PH_3和CA交替处理(ALT-s)的死亡率相对PH_3或CA单独处理显著增加,处理5次后(3次PH_3处理,2次CA处理),种群死亡率达100%。

表5 每次暴露于CA和PH_3后种群中嗜卷书虱的平均数量±标准误

处理*	暴露时间/周					
	1(1)	2(3)	3(5)	4(7)	5(9)	6(11)
CA	25.0±2.64b	278.7±15.94b	121.7±6.77b	169.7±5.61c	141.3±11.57c	129.0±5.51c
PH_3	12.4±1.03c	172.1±3.56c	87.3±5.93c	245.2±11.25b	418.4±22.56b	372.0±12.06b
交替处理	12.4±1.03c	49.7±1.33 d	37.2±1.62 d	45.3±6.75 d	0.0±0.00 d	0.0±0.00 d
对照	30.0±0.00a	385.0±25.11a	422.7±12.20a	1 294.3±51.45a	1 379.0±92.44a	1 443.7±55.51a
F ($df=3$,8)	51.62	134.58	73.03	369.53	388.12	633.41
P	0.000 1	0.000 1	0.000 1	0.000 1	0.000 1	0.000 1

注:*每个种群开始时包含30只新羽化的成虫,并以2周的间隔进行处理。CA为35%CO_2、1%O_2和64%N_2,在第1周处理4 h,然后在随后的每次处理中增加2 h。PH_3每次以12 μg/L的浓度处理,持续24 h。交替处理是在第1、5和9周使用PH_3,在第3、7和11周使用CA。对照组不暴露于混合气体或PH_3。每种处理重复三次。

3 讨论

1)嗜卷书虱的卵、成虫及若虫对PH_3的敏感性有很大差异。24 h的熏蒸处理,卵的LC_{50}为0.137 mg/L,是一龄若虫的31.97倍,是成虫的6.71倍(表2、表3)。采用25 μg/L的PH_3进行24 h熏蒸处理,间隔10 d后再分别进行第2次和第3次熏蒸处理,可以完全控制嗜卷书虱的发生。一般认为0.25 mg/L是粮仓中的合适使用浓度,对大多数昆虫在120 h的控制浓度一般仅为

0.20 mg/L。本研究使用0.25 mg/L的浓度熏蒸处理120 h,对嗜卷书虱卵的致死率仅为91.6%(表1),LC_{99}为0.587 mg/L(表2),而这一浓度可以将嗜卷书虱的成虫和若虫全部杀死(表4)。此外,嗜卷书虱卵的发育天数为8 d左右,卵对PH_3又有较强的忍耐力,实仓熏蒸时,一般采用可使成虫致死的浓度。而且采用一定浓度的PH_3熏蒸,5 d后PH_3的实际浓度就会大大降低。因此,正常的熏蒸处理,很难杀死嗜卷书虱的卵。Pike的研究表明,对嗜虫书虱的卵来说,72 h的LC_{99}为7.01 mg/L,这在一般熏蒸处理的条件下是很难达到的。这就可以解释为什么在粮仓中采用PH_3熏蒸处理难以有效治理书虱的问题。

2)采用间歇熏蒸和气调与熏蒸交替处理的方法可以解决书虱防治困难的问题。连续进行3次熏蒸可以达到100%的致死效果(表4),而气调与PH_3熏蒸的交替处理,在剂量较低的情况下,也可以达到很好的效果,这种情况与Ding等研究的气调和敌敌畏交替处理对嗜卷书虱抗性增长与种群发展的影响有相同之处。因此,在书虱治理的过程中,应采用综合治理的措施对其进行控制,如采用间歇熏蒸处理,以及气调与药剂(包括PH_3和敌敌畏)交替处理的方法或者CO_2和PH_3的混合熏蒸,同时还要配合调节仓内的生态环境等。只有这样,才能有效地控制书虱的发生与危害。

3)PH_3熏蒸对昆虫的作用机理目前还没有明确的结论,一般认为它同氢氰酸的生理效应比较相似,也是抑制呼吸链中最后的一个酶,即细胞色素C氧化酶。该酶一旦被抑制,则引起昆虫组织中三磷酸腺苷(ATP)的衰竭,由此而终止了昆虫对氧的利用和能量的产生。所以,从生理学来看,PH_3是昆虫的呼吸抑制剂。Price曾证明,氧的存在是PH_3对昆虫毒性作用所必需的,因此,可以认为,氧是PH_3毒性作用的条件;昆虫对氧的利用的终止,是毒性作用的结果。所以,PH_3对嗜卷书虱的作用与嗜卷书虱本身的生理活动状态密切相关,不同虫态对PH_3敏感性的差异表明,新陈代谢作用与熏蒸毒剂的致命性有密切的关系。此外,气调措施可以改变昆虫的需氧条件,主要影响昆虫的需氧代谢,气调与PH_3的作用机理是有差异的,因此,采用气调与PH_3交替处理的办法,不仅可以延缓嗜卷书虱抗性的发展,而且可以很好地控制该虫的种群数量。

原文刊载于《农药学学报》,2003年第3期

CO_2 和溴氰菊酯在不同温度下对嗜卷书虱毒性的相互影响

吴仕源,王进军,赵志模

摘要:采用三元一次正交组合实验设计,在18%O_2浓度下研究了CO_2浓度25%~35%(X_2)、溴氰菊酯3.54×10^{-4}~7.07×10^{-4} μg/cm^2(X_3)和温度20~30 ℃(X_1)不同水平组合对嗜卷书虱(*Liposcelis bostrychophila*)的急性致死作用。结果表明,CO_2浓度是影响嗜卷书虱半数致死时间(LT_{50})变化的主要因素,其次是溴氰菊酯,然后是温度因素。3因素各编码水平对嗜卷书虱的半数致死时间(LT_{50},Y)的回归关系式为:$Y=6.547-0.701X_1-2.181X_2-1.206X_3+0.648X_1X_2$。在高浓度的$CO_2$中$CO_2$是导致嗜卷书虱死亡的主要因素。在低浓度$CO_2$中温度升高对嗜卷书虱$LT_{50}$影响显著,随着$CO_2$浓度升高温度因素的作用逐渐减小,$CO_2$和温度之间存在着显著的交互作用;在一定浓度的气调中增加适当浓度的溴氰菊酯就可有效地控制该虫的危害。

关键词:嗜卷书虱;CO_2;溴氰菊酯;温度;半数致死时间(LT_{50})

The Acute Toxicity of Carbon Dioxide and Deltamethrin on *Liposcelis bostrychophila* (Badonnel) at Different Temperatures

Wu Shiyuan, Wang Jinjun, Zhao Zhimo

Abstract: This paper deals with the acute toxicity of the different level combination of temperature (X_1, 20—30 ℃), carbon dioxide (X_2, 25%—35%) and deltamethrin (X_3, 3.54×10^{-4}—7.07×10^{-4} μg/cm^2) with orthogonal combination design on the psocid, *Liposcelis bostrychophila*. The results showed that high CO_2 concentration was the key factor resulting in the mortality of the psocids, deltamethrin was the less important one, and temperature was the least. The regression equation between the half lethal time (LT_{50}, Y) and the factors was obtained as follows: $Y=6.547-0.701X_1-2.181X_2-1.206X_3+0.648X_1X_2$. The interaction between CO_2 and temperature significantly affected the LT_{50} of the psocids. The LT_{50} would become shorter with the increase of temperature under low CO_2 concentration, obviously. However, under higher CO_2 concentration, the influence of temperature was not significant.

Considerably higher CO_2 concentration combined with certain deltamethrin concentration was effective for controlling psocids.

Key words: *Liposcelis bostrychophila*; CO_2; deltamethrin; temperature; LT_{50}

0 前言

嗜卷书虱[(*Liposcelis bostrychophila*(Badonnel)]属啮虫目(Psocoptera),书虱科(Liposcelididae),是仓库内广泛分布的害虫之一。长期以来一直被认为是次要性害虫,直至二十世纪八十年代其危害性才逐渐受到人们的重视。特别是在目前广泛采用的"三低""双低"储粮仓库中嗜卷书虱已经上升为优势种群,尚缺乏有效的防治措施。本研究选用目前粮仓中使用较普遍的杀虫剂——溴氰菊酯,研究了它在不同温度下配合高CO_2气调措施对嗜卷书虱的急性致死作用,旨在为综合治理该害虫提供科学依据。

1 试验材料和方法

1.1 试验材料及设备

嗜卷书虱:西南农业大学植保系仓虫生态室多年室内饲养的纯系,用羽化后3~6 d的成虫作为供试虫源。

2.5%溴氰菊酯乳油:法国罗素优克福公司生产。用蒸馏水分别稀释为80,120,160 μg/mL的浓度备用。

气调设备:西南农业大学应用昆虫及螨类生态研究室设计组装的气体浓度控制仪,CO_2、O_2、N_2(纯度:CO_2>99.0%,O_2>99.2%,N_2>99.9%)分别从钢瓶中流出,经减压阀,稳压阀,流量控制阀,转子流量计和气体混合瓶后,按不同的气体组配调节进入试验处理容器,其中N_2作为平衡气体,O_2恒定为18%。

1.2 试验方法

1.2.1 处理方法

温度(X_1)、CO_2(X_2)和溴氰菊酯(X_3)按三元一次正交组合设计,各因子水平及编码见表1,实验结构矩阵见表2。

表1 试验水平

因子		编码		
		-1	0	1
温度/℃	(X_1)	20	25	30
CO_2/%	(X_2)	25	30	35
溴氰菊酯/($\mu g/cm^2$)	(X_3)	3.54×10^{-4}	5.31×10^{-4}	7.07×10^{-4}

把直径为2.4 cm的圆形滤纸片放入内径为2.5 cm,高为1.0 cm的圆形塑料盒中,用滴定管将0.02 mL配制好的溴氰菊酯药液滴在滤纸片上,10 min后,各试验盒接入不少于50头书虱成虫,再盖上160目尼龙丝网做成的盒盖。塑料盒悬挂于容积1 000 mL的广口试验瓶中,试验瓶底部用饱和食盐水调节湿度至75%~85%RH。

将CO_2、N_2和O_2配合成不同的处理浓度,以250 mL/min流速通过试验瓶30 min后,把试验瓶密封放入恒温室,处理一定时间后将试验盒放置在自然大气条件下12 h,然后观察记录供试虫的死亡情况。每处理重复3次,同时设置对照(试验盒的滤纸上滴0.02 mL清水,试验瓶内保持自然大气)。

1.2.2 处理时间

在初试的基础上,保持死亡率在10%~90%范围内,选取一系列适当的时间处理试虫。每编号至少进行5个不同的有效时间处理。

1.2.3 LT_{50}的计算

取各处理死亡率在10%~90%之间所需的时间值,并将其转换为常用对数,死亡率用Abbot公式校正后转换成机率值,利用最小二乘法获得二者间的回归关系式,在机率值为5.0时,计算所得到的Y值的反对数即为LT_{50}。

表2 试验结构矩阵及结果

编号No.	温度(X_1)	CO_2(X_2)	溴氰菊酯(X_3)	截距a	斜率b	Y	$LT_{50}(10^Y)$/h
1	−1	−1	−1	1.460	3.375	1.049	11.2
2	−1	−1	1	1.151	4.211	0.914	8.2
3	−1	1	−1	0.994	5.730	0.699	5.0
4	−1	1	1	1.963	6.013	0.505	3.2
5	1	−1	−1	2.315	2.890	0.929	8.5
6	1	−1	1	1.271	5.039	0.740	5.5
7	1	1	−1	0.927	5.981	0.608	4.8
8	1	1	1	2.017	6.457	0.462	2.9
9	0	0	0	1.745	3.825	0.851	7.1
10	0	0	0	1.667	3.736	0.892	7.8
11	0	0	0	1.548	4.210	0.820	6.6
12	0	0	0	1.733	3.812	0.857	7.2
13	0	0	0	2.004	3.571	0.819	6.9
14	0	0	0	1.950	3.692	0.826	6.7

2 结果分析

2.1 温度、CO_2和溴氰菊酯与LT_{50}间总回归关系

表2表明了温度(X_1)、CO_2(X_2)和溴氰菊酯(X_3)不同水平组合下对嗜卷书虱的半数致死时间(LT_{50})。对嗜卷书虱LT_{50}(Y)与温度(X_1)、CO_2(X_2)和溴氰菊酯(X_3)间进行一次正交组合的回归分析,去除统计检验不显著的回归项得回归模型:

$Y=6.542\ 8-0.737\ 5X_1-2.187\ 5X_2-1.212\ 5X_3+0.612\ 5X_1X_2$ ($R=0.930\ 3^{**}$)

经方差分析检验,3因素与嗜卷书虱LT_{50}间总回归关系达到极显著($F=30.03$)。

2.2 单因素对嗜卷书虱半数致死时间(LT_{50})的影响

从回归模型中的回归系数可以看出，$CO_2(X_2)$对LT_{50}贡献最大，其次是溴氰菊酯(X_3)和温度(X_1)，然后是$CO_2(X_2)$和温度(X_1)的交互作用，表明它们是影响LT_{50}变化的主要因素。

由于本试验具有正交性，因此固定两个变量的水平为0可得到温度、CO_2和溴氰菊酯与嗜卷书虱LT_{50}间的单因素直线回归模型（图1）。由图1可知，CO_2浓度与LT_{50}回归直线的斜率最大，说明CO_2浓度是影响嗜卷书虱半数致死时间(LT_{50})变化的主要因素，其次是溴氰菊酯的单位含量($\mu g/cm^2$)，然后是温度水平。

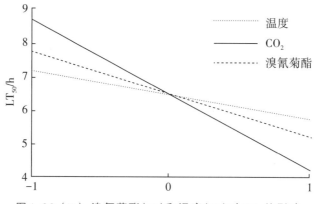

图1 $CO_2(X_2)$、溴氰菊酯(X_3)和温度(X_1)对LT_{50}的影响

2.3 双因素对嗜卷书虱半数致死时间(LT_{50})的影响

2.3.1 CO_2和温度对嗜卷书虱半数致死时间(LT_{50})的影响

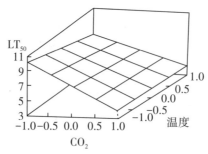

图2 $CO_2(X_2)$和温度(X_1)对LT_{50}的影响

从图2中可以看出,在不同温度水平上随着CO_2浓度的增高,嗜卷书虱半数致死时间LT_{50}降低,其斜率在20 ℃水平上最大($b=-2.800$),35 ℃水平上最小($b=-1.570$),统计分析表明两者之间的交互作用达到极显著水平。从图2中还可以看出,在不同CO_2水平上温度的升高对嗜卷书虱LT_{50}的影响,在低CO_2(25%)水平上随着温度升高LT_{50}随着降低,其斜率为-1.390;但是高CO_2(35%)水平下LT_{50}直线斜率为-0.165,表明在高CO_2中温度对嗜卷书虱LT_{50}的影响较小。在低CO_2中温度和CO_2的交互作用对嗜卷书虱LT_{50}的影响显著。这是因为温度升高可以促使虫体酶活性增强,新陈代谢旺盛,呼吸速率升高,从而导致吸入过量CO_2而死亡。然而,在CO_2浓度较高的情况下,CO_2浓度足以导致虫体死亡,因而掩盖了温度对加速虫体死亡的增效作用。

2.3.2 CO_2和溴氰菊酯对嗜卷书虱半数致死时间(LT_{50})的影响

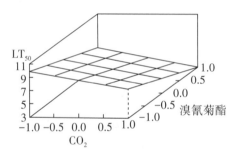

图3 CO_2(X_2)和溴氰菊酯(X_3)对LT_{50}的影响

从图3看出,CO_2浓度同溴氰菊酯每平方厘米含量间有较显著的相互增效作用。在不同CO_2水平上随着溴氰菊酯单位含量的增高,嗜卷书虱LT_{50}降低;同时,在不同的溴氰菊酯含量水平上,随着CO_2浓度升高,嗜卷书虱LT_{50}也随着降低。另外,还可以看出溴氰菊酯对LT_{50}影响的变化率比CO_2大,而且统计分析表明两者之间的交互作用没有达到显著水平,这说明在CO_2气调中增加适当的化学药剂就可以加速虫体的死亡。

2.3.3 溴氰菊酯和温度对嗜卷书虱半数致死时间LT_{50}的影响

从图4看出,在不同溴氰菊酯单位含量水平上,温度对嗜卷书虱LT_{50}的影响力较小,表现在其直线斜率小。在不同温度水平上溴氰菊酯随着单位含量升高，LT_{50}随着降低。但这种交互作用也不太显著。

综上所述，影响嗜卷书虱半数致死LT_{50}的主要因素是CO_2，其次是溴氰菊酯，然后是温度因素；CO_2和温度之间交互作用效果明显。

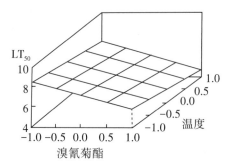

图4 溴氰菊酯(X_3)和温度(X_1)对LT_{50}的影响

3 结论

嗜卷书虱是我国粮仓内广泛分布的主要害虫之一。特别是我国南方广大地区，由于气候温暖湿润，书虱大量繁殖，加之防治不易，已经成为目前粮食储藏中的一个重要问题。国外的一些研究表明，嗜卷书虱不仅对溴氰菊酯等合成除虫菊酯有忍耐力，而且它对含高浓度CO_2的大气具有忍耐力。因此，防治书虱危害不能单靠一种方法，要遵循"以防为主，综合治理"的方针。根据本研究成果，应以CO_2气体为主，并辅之以适当浓度的化学药剂，再配合适宜的温度措施，才能安全有效地控制嗜卷书虱的危害。

原文刊载于《中国粮油学报》，1997年第6期

Accumulation and Utilization of Triacylglycerol and Polysaccharides in *Liposcelis bostrychophila* (Psocoptera, Liposcelididae) Selected for Resistance to Carbon Dioxide

Wang Jinjun, Zhao Zhimo

Abstract: Two populations of the psocid, *Liposcelis bostrychophila* Badonnel, were exposed to two CO_2-enriched atmospheres (35% CO_2+21% O_2, and 55% CO_2+ 21% O_2, balance N_2) for 30 generations. Controls were reared in normal atmospheres. The reserves of triacylglycerol and polysaccharides were evaluated in adults of the two experimental and the control populations in generations F_{15} and F_{30}. The utilization rate of triacylglycerol and polysaccharides in the CO_2-enriched atmospheres were also determined in generation F_{30}. The results indicated that the reserves of triacylglycerol and polysaccharides increased significantly during selection for CO_2 resistance; the higher the resistance level, the greater the reserves. Exposure of these populations to controlled atmosphere was associated with a steady utilization of the reserves. By contrast, the unselected population responded to controlled atmospheres by accelerated utilization of triacylglycerol and polysaccharides. Comparison of the utilization rates during CO_2 exposure showed that triacylglycerol is the main energy source, and polysaccharides contribute to metabolic energy supply only to a small extent.

1 Introduction

One of the most widely spread members of the order Psocoptera is a minute apterous species *Liposcelis bostrychophila* Badonnel, commonly, but erroneously, known as booklouse. *L. bostrychophila* is commonly found in various processed and unprocessed dry foods in households, granaries and warehouses. Apart from causing measurable damage to stored grains, infestations of this psocid can also cause health problems among storage and warehouse workers. This insect feeds continually and reproduces rapidly by obligatory thelytokous parthenogenesis. Outbreaks of *L. bostrychophila*, together with *Liposcelis entomophila* Enderlein, and *Li-*

poscelis paeta Pearman, have been reported in humid tropic countries such as Indonesia, Malaysia, Singapore, The Philippines, Thailand, The People's Republic of China and India. In China, *L. bostrychophila* has become one of the dominant pest species for sealed grain storage. Routine fumigation of warehouses and storage facilities with methyl bromide or using insecticides have failed to control this pest, that can readily reinfest grain in storage after fumigation or insecticides treatments. In addition, rapid development of resistance to chemical and physical treatments has also been reported for *L. bostrychophila*.

Controlled atmospheres (CA) are alternative treatments to the use of methyl bromide for post-harvest insect control. CA with elevated CO_2, depleted oxygen levels, or a combination of both have been tested and showed promise for control of stored product pests. The development of insect resistance to most commonly used fumigants such as phosphine, methyl bromide and ethylene dibromide has been well documented. Our previous study showed that two laboratory populations of the psocid *L. bostrychophila* were selected for resistance to hypercarbia. Similar phenomenon was reported for other insect species including *Sitophilus granarius* (L.), *Sitophilus oryzae* L., and *Tribolium castaneum* (Herbst). Donahaye stated that the development of resistance to hypoxia or to hypercarbia through selection may be accompanied by physiological and biochemical changes. These changes may enhance insect ability to survive environmental stress. However, there have been few attempts to explain insect resistances to hypercarbia or hypoxia on the basis of physiological or biochemical data. Lack of such knowledge has rendered the development of CA treatments costly and time consuming. It is known that insect survival time depends on the energy requirements of the particular organs of the individual concerned during CA exposure. Triacylglycerols (TG) and polysaccharides are generally considered as the principal energy reserves of insects. Most recently, Donahaye and Navarro reported that the quantities of energy reserves and their utilization in *T. castaneum* were closely correlated to CA resistance. In this study, the changes of reserves and utilization of TG and polysaccharides during resistance selection, and during CA exposures were evaluated in three populations of

Liposcelis bostrychophila. We attempted to elucidate the effects of CO_2 on *L. bostrychophila* and to determine the function of energy metabolism in relation to the development of resistance.

2 Materials and methods

2.1 Test insect and rearing conditions

Stock colonies of *L. bostrychophila* originated from nymphs collected in a wheat warehouse in Chongqing, P.R. China in 1990. These colonies were reared on a diet consisting of whole wheat flour, skimmed milk, and yeast powder (10:1:1) in an air-conditioned room at (28±1) ℃, and a scotoperiod of 24 h. Cultures were set up in glass bottles (250 mL) with a nylon screen cover and kept in desiccators (5 000 mL), in which the humidity was maintained at 75%—80% with saturated NaCl solution. After many generations, insects from the stock colony were used for the tests. The identity of *L. bostrychophila* was confirmed by F.S. Li (China Agricultural University, Beijing, P.R. China). Voucher specimens were deposited at the insect collection of Southwest Agricultural University, Chongqing, P.R. China.

Three populations of *L. bostrychophila* were used in this experiment. One population (S1) has been selected to 35% CO_2, 21% O_2, and the other (S2) to 55% CO_2, 21% O_2, both balanced by N_2. A third population, the original unselected laboratory population sensitive to CA, was also employed. The resistance factors of S1 were 3.2- and 4.6-fold, and S2 were 3.6- and 5.3-fold at the generation 15 and 30, respectively. The detailed information on selection procedure and resistance development were reported by Wang et al.

2.2 Triacylglycerol measurements

Four groups of 100 newly emerged adults taken from cultures of all three populations were analysed. Total lipid extraction was carried out by the method of Bligh and Dyer. The assay methods for TG measurement have been previously outlined by Bucolo and David. The TG concentration was expressed in microgram trioleoylglycerol per insect.

TG reserve was measured at the 15th and 30th generations. TG utilization during exposure to CAs was performed at the 30th generation. The exposure experiments were carried out in a controlled atmosphere unit capable of generating a mixture of three gases (CO_2, O_2 and N_2) in the range of 0%—100%. All gas concentration measurements were carried out with the same gas chromatography apparatus at 2 h intervals as previously described by Wang et al. The apparatus was kept in a temperature-controlled room at (28±1) ℃. The mixed and humified gases were flushed continuously into the exposure units of the insect chamber in order to prevent changes of gas composition caused by metabolic activities of the test insects. Adults 2—5 days old were taken from cultures of the 30th generation of all three populations. These insects were exposed to 35% CO_2, 1% O_2 and balance N_2 without food for the following exposure times: 0, 2, 4, 6, and 8 h. At the end of each exposure time, the insects were removed from the exposure apparatus and TG concentration was measured. By subtraction of recorded TG concentrations after exposure from the TG concentration recorded before exposure, the utilization rate of TG was determined for each exposure period.

2.3 Polysaccharide measurements

The reserves and utilization of polysaccharides were determined at the same time schedule as TG measurements. The polysaccharide assay was identical to that of Keppler and Decker without distinction between glycogen, amylose and amylopectin, and of concentrations of glucose, whereas chitin as a source of polysaccharide was not included in the test. Concentration of polysaccharides is expressed in microgram per insect.

Differences of TG and polysaccharides concentrations among different populations were subjected to analysis of variances (ANOVA). General linear model procedure was used and mean values were separated by Fisher protected least significant difference (LSD) test when significant F-value was obtained ($P<0.05$). The percentage of utilization rate was transformed to the arcsin square-root before analysis to stabilize error variance.

3 Results

3.1 Triacylglycerol measurements

The TG measurements for the three populations are presented in table 1. The average initial TG concentration for the control population was 2.65 μg TG per specimen, and it remained unchanged in generations F_{15} and F_{30} ($F=0.49$; $d.f.=2, 9$; ns). The TG concentrations for S1 and S2 increased significantly during selection (S1: $F=1\ 654.3$; S2: $F=1\ 923.1$; $d.f.=2, 9$; $P<0.001$). Among the three populations, the differences in TG concentrations were significant at the generations F_{15} and F_{30} ($P<0.001$; table 1), whereas they were same before selection ($P\geqslant0.05$). By generation F_{30}, the TG concentrations in S1 and S2 were 2.30- and 2.75-fold higher than that of the control population, respectively. Between two selected populations, the TG concentration in S2 was significantly higher than that in S1 at both F_{15} and F_{30} (Table 1).

Table1. Comparison of triacylglycerol concentrations (μg TG per psocid) in adults of three populations of *Liposcelis bostrychophila*

Population	Initial values	Generation F_{15}	Generation F_{30}
Control population	2.65±0.07a (A)	2.65±0.05a (A)	2.62±0.13a (A)
S1	2.62±0.01a (A)	4.65±0.18b (B)	6.02±0.09b (C)
S2	2.64±0.04a (A)	5.16±0.11c (B)	7.20±0.23c (C)
F ($d.f.=2, 9$)	0.74	1 716.20	1 799.90
P	ns	0.001	0.001

Each value represents the mean (±SD) of four replicates; each replicate consists of 50 adults. Mean values within a column (row) followed by same lower (upper) case letters are not significantly different ($P\geqslant0.05$). Populations S1 and S2 are described in 'Materials and methods'.

TG utilization during exposure of the three populations to the controlled atmosphere is presented in Figure 1. As mentioned above, TG concentrations were significantly different ($P<0.05$) between the three populations. During exposure, the TG utilization rate was significantly higher in the control population than in the two CA selected populations. After 8 h exposure, the TG utilization rates for the control, S1, and S2 were 43.0%, 21.2% and 16.9%, respectively. Only some individuals were able to recover after 8 h exposure in the control population, whereas more than 75% individuals survived in both selected populations.

3.2 Polysaccharide measurements

The polysaccharides concentration in the control population remained stable throughout the experiment ($F=0.32$; $d.f.=2, 9$; ns) and the mean value was 1.19 μg per psocid (table 2). The polysaccharide concentrations in two selected populations increased significantly up to 1.44- and 1.67-fold by generation F_{30} for S1 and S2, respectively ($P<0.001$).

The utilization rates of polysaccharides in all three populations increased over the exposure time (Figure 2). The utilization rate of polysaccharides in the control population was higher than in the two selected populations. However, the utilization rates of polysaccharides were rather low in all three populations. After 8 h exposure, the utilization rate of polysaccharides in the control population was only 11.9%, in spite of very high mortality. Two selected populations utilized 8.6 and 7.3% of their initial reserves of polysaccharides after 8 h exposure.

4 Discussion

Not all aspects of the mode of action of CO_2 on insects have been elucidated. Little information is available in the literature on lipid and carbohydrate metabolism under the stress caused by hypoxia or hypercarbia. We observed that the control population of *L. bostrychophila* exposed to CO_2-enriched atmosphere lost body water more rapidly than the selected populations. The insects eventually died of desiccation. Similar phenomenon of water loss during exposure was reported for *Tribolium castaneum*, and for *Platynota stultana* Walsingham. The survival of *Liposcelis bostrychophila* depends on the water content. Normally, water makes up about 66% of the body mass in adult *L. bostrychophila* that can tolerate a loss of half of this amount before succumbing. Whilst some water is obtained from the food moisture as well as from the metabolic breakdown of carbohydrates, much comes from water vapour absorption from the atmosphere. Psocids are highly unusual in this respect.

Table2. Comparison of polysaccharide concentrations (μg per psocid) in adults of three populations of *L. bostrychophila*

Population	Initial values	Generation F_{15}	Generation F_{30}
Control population	1.17±0.05a (A)	1.20±0.01a (A)	1.20±0.01a (A)
S1	1.20±0.01a (A)	1.53±0.04b (B)	1.73±0.04b (C)
S2	1.19±0.03a (A)	1.80±0.01c (B)	2.00±0.03c (C)
F (d.f.=2, 9)	0.69	816.4	839.7
P	ns	0.001	0.001

Each value represents the mean (±SD) of four replicates; each replicate consists of 50 adults. Mean values within a column (row) followed by same lower (upper) case letters are not significantly different ($P \geq 0.05$). Populations S1 and S2 are described in 'Materials and methods'.

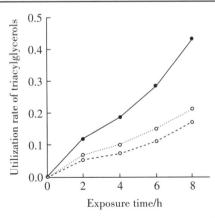

Figure1. Utilization rate of triacylglycerols during controlled atmosphere (CA) exposure for control population (solid dot and solid line), S1 (circle and dashed line) and S2 (hexagons and dotted line) at generation F_{30}

The function of TG in the conservation of water balance in insects have been well documented. Donahaye reported that the difference in TG utilization between the unselected and CA-selected populations corresponds to their different abilities to maintain water balance during exposure to enhanced CO_2 concentration. Friedlander and Navarro found that TG metabolism was not sufficient to compensate for water losses in *Ephestia cautella* (Walker) pupae under conditions of high CO_2 concentration and low ambient humidity. In this study, in order to ensure that the insects utilized only their initial energy reserves, they were exposed to the CAs without food. The results showed that the CA-selected populations had significantly greater TG reserves. It can be concluded that the selection for CO_2 resistance was actually selection for a certain type of TG metabolism. Higher resistance level was associated with accumulation of higher TG reserves and slower rate of TG utilization. The present study also showed that exposure to CA was accompanied by rapid utilization of TG, particularly in the non-adapted control population. The mortality of control population was also significantly higher than that of two selected populations. This implied that the ability to control the loss of metabolic water through steady and moderate utilization of TG prolongs survival of the selected population during CO_2 exposure. This is consistent with the results reported by Friedlander et al. They suggested that elevated CO_2 migh affect energy use in several ways. It reduces NADPH production and inhibits the biosynthesis of glutathione. Zhou et al. reported that high CO_2 also affect the ATP production by the influence on insects' membrane systems. Hoyle demonstrated the action of CO_2 on the spiracular muscle. The extent of spiracular control during expoure to CAs may also influence the ability of insect to resist desiccation, and this has been shown to be a contributing factor to mortality under CA exposure. However, this aspect in *L. bostrychophila* remained unknown and needs further investigation.

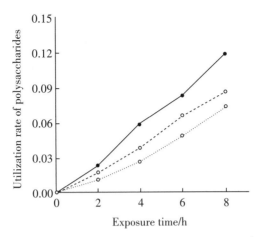

Figure 2. Utilization rate of polysaccharides during controlled atmosphere (CA) exposure for control population (solid dot and solid line), S1 (circle and dashed line) and S2 (hexagons and dotted line) at generation F_{30}

Friedlander and Navarro reported that exposure of *Ephestia cautella* pupae to hypercarbia increased glycogen utilization. Glycogen forms one of the major carbohydrate energy reserves in insects, and may account for up to 2% of the wet weight of insect. It has been proposed that the effects of hypercarbia on insects do not exclude the effects of hypoxia because high CO_2 can prevent insects from using oxygen. Animals mainly use two strategies to cope up with hypoxia: anaerobic metabolism and metabolic arrest. Anaerobic metabolism can temporarily compensate for energy insufficiency of oxidative phosphorylation. However, this strategy would require very high rates of glycolysis and thus lead to rapid exhaustion of carbohydrate reserves while toxic end products accumulate. However, the present study showed that all populations utilized polysaccharides very slowly. This suggested that the glycolytic pathway is not of major importance during exposure of *Liposcelis bostrychophila* to CAs.

In general, the advantage of CA-selected populations in TG and polysaccharides reserves over selection, and slower utilization rate during exposure is evident. The higher quantities of TG reserves in the CA-selected populations and their lower rate of utilization, enable them to survive longer under CA exposure, and this might be an important mechanism of resistance to CA in *L. bostrychophila*.

Nonetheless, further investigation on other physiological, together with biochemical and biological aspects are needed to clarify the resistance mechanisms.

Original article was published in *Journal of Applied Entomology*, volume 127, 2003.

Development and Reproduction of Three *Euseius* (Acari：Phytoseiidae) Species in the Presence and Absence of Supplementary Foods

Zhao Zhimo, McMurtry James Allen

Abstract：Development duration, mortality and oviposition rate of *Euseius tularensis* (Congdon), *E. stipulatus* (Athias-Henriot) and *E. hibisci* (Chant) were determined on three basic foods (pollen of iceplant, Pacific spider mite and citrus red mite) in the presence and absence of supplementary foods (honeydew from aphids and whiteflies). The studies were conducted in the laboratory at a constant temperature of 25 ℃. The mean duration of each development stage and total development time did not differ significantly between treatments with and without supplementary food, regardless of species and basic foods. The mortality of the immature stages of the three *Euseius* species was generally higher in the absence of supplementary foods than in the presence of these foods, but differences were not significant ($P=0.05$). Oviposition rates generally were significantly higher in the presence of supplementary foods. The basic food was the most important factor affecting development, survival and oviposition rate; the most favorable food was pollen and the least favorable was citrus red mite. The development and oviposition rates differed somewhat among the three *Euseius* specieson the same food or food combination.

Introduction

The phytoseiid mites *Euseius hibisci* (Chant), *E. tularensis* (Congdon) and *E. stipulatus* (Athias-Henriot) are important natural enemies on citrus and avocado in southern California. In groves where there is minimal pesticide interference, these acarine predators generally can be important mortality factors in biological control of the citrus red mite, *Panonychus citri* (McGregor) on citrus, *Oligonychus punicae* (Hirst) and *Eotetranychus sexmaculatus* (Riley) on avocado, and the citrus thrips, *Scirtothrips citri* (Moulton), on citrus. Recent evidence obtained by morphological and hybridization studies demonstrated that the mite referred to as *Euseius hibisci* in earlier papers involves two sibling species. The first species,

E. hibisci, tends to be limited to coastal areas of southern California and occurs on both citrus and avocado. The second species, *E. tularensis*, is found on citrus, especially in the inland areas of southern California and the San Joaquin Valley, and rarely is found on avocado. *Euseius stipulatus* was introduced from the Mediterranean region and is established on citrus in coastal areas of southern California where it has displaced *E. hibisci* on citrus in many orchards in San Diego, Ventura and Santa Barbara counties.

These three species of phytoseiid mites feed not only on tetranychid mites and thrips, but also on pollen and honeydew. Some species of *Euseius* can complete their life-cycle and reproduce more rapidly on pollen than on tetranychid mites. McMurtry and Scriven also observed that *E. hibisci* from coastal southern California laid more eggs when feeding on tetranychid mites in combination with an agar-base medium containing yeast hydrolysate than when feeding on mite prey alone. McMurtry and Scriven found that female *E. hibisci* fed a combination of spider mites and honeydew during development had a shorter mean time to oviposition compared with those fed mites alone, and that the phytoseiids could survive for more than one month on honeydew alone. This indicated that supplementary foods such as honeydew and yeast probably play an important role in the diet of these predaceous mites.

It was felt that further studies of feeding habits of the three *Euseius* species might provide a better understanding of some of the factors affecting their abundance in the field. The objective of this study was to evaluate and compare the development, mortality and reproduction of these species on different basic foods in the presence and absence of supplementary foods.

Materials and methods

The test arenas consisted of excised 'Valencia' orange leaves placed upper-surface-down on 12-mm-thick, foam mats (17 cm×17 cm) in stainless-steel pans (20 cm×20 cm) with water. Each leaf arena (4 cm×4 cm) was surrounded by a 1.5-cm-wide strip of Cellucotton®. Each steel pan contained 9 such leaf arenas. The Cellucotton strips, saturated with water, served as a barrier to prevent the predaceous mites from escaping from the arenas and to keep the leaves moist. Whenever a leaf began to deteriorate, it was replaced with another fresh leaf. All experiments were conducted in small, modified refrigerators at (25 ± 0.5) ℃, 40%—70% relative humidity (RH), 12 hours of light and 12 hours of darkness. None of the species is known to have a diapause.

The basic foods tested for the three *Euseius* species included pollen of the iceplant *Malephora crocea* (Jacq.), eggs of the Pacific mite *Tetranychus pacificus* McGregor, extracted from glasshouse-grown lima bean plants by the method described by Scriven and McMurtry (the eggs began to hatch about one day after being placed into test arenas), and eggs of the citrus red mite *Panonychus citri* (McGregor). Three days before the test began, approximately 10 adult females of citrus red mite were transferred from field-collected citrus leaves to each test arena. Consequently, when the predaceous mites were introduced to test arenas, numerous eggs and a few larvae of citrus red mite were available as food. Five additional adult females were transferred daily to each test arena during the experiment.

The supplementary foods tested were honeydew from the bean aphid *Aphis fabae* Scopoli, and from the woolly whitefly *Aleurothrixus floccosus* (Maskell). The aphid honeydew was obtained by placing 15 mm×15 mm cover glasses for 24 h under broad-bean plants heavily infested with aphids in the greenhouse. The honeydew of whiteflies was obtained by collecting a large number of orange leaves infested with woolly whiteflies. The honeydew was transferred from the whitefly colonies drop by drop with a pin onto cover glasses. At least 10 drops of honeydew were placed on each cover glass. One of these cover glasses then was placed on each test arena. When honeydew became extensively covered by growth of microorganisms (ca. 7—10 days), it was replaced with another coverslip with fresh honeydew.

Nine experimental treatmentswere setup for each of the three *Euseius* species:

(1) iceplant pollen alone (P); (2) pollen plus aphid honeydew (P+AH); (3) pollen plus whitefly honeydew (P+WH); (4) Pacific spider mite (PM) only; (5) PM+AH; (6) PM+WH; (7) citrus red mite (RM) alone; (8) RM+AH; and (9) RM+WH. Six replicates, each containing 5 eggs of the phytoseiid species on a test arena, were employed for each experiment treatment.

The phytoseiids were obtained from insectary cultures. Samples of 20—30 adult females from each culture were mounted and identified in order to confirm that cultures were not contaminated with other phytoseiid species. For development studies, eggs were obtained by placing clumps of cotton on tiles of rearing units of the stock culture. After 24 h, the clumps were examined and eggs of the predaceous mites that had been oviposited thereon were collected and placed in test arenas. Each species was studied separately. To determine the duration of development stages, observations were made every 12 h until all individuals reached the adult stage.

To determine oviposition rates, newly mature females and males were taken from the development experiment and introduced onto new test arenas, each containing two females and two males as a replicate. Each treatment was replicated three times. Eggs laid by these females were counted and removed daily for 11—12 days. The mean number of eggs laid per female per day was compared by Duncan's multiple-range test at the end of the experiment. To determine the approximate duration of F_1 eggs, 10 eggs oviposited on the same day were taken from each treatment and placed in a new test arena. These eggs were observed every 12 h until all eggs hatched.

Results and discussion

All three *Euseius* species fed and developed to maturity on all foods or food combinations, but their development rates varied with species and foods, especially basic foods. The development of the three species on three different basic foods in the presence and absence of supplementary foods is shown in Table 1. The differences in the development duration of each mite species on different basic foods (averages of the three treatments for each basic food) are shown in Figure 1.

The mean duration of each development stage and total time for immature development were not significantly affected by the presence of supplementary foods, regardless of *Euseius* species and basic food (Table 1). Basic food was an important factor affecting the rate of development of these mites. The fastest rate occurred when the mites were fed on pollen. These results are consistent with those of McMurtry and Scriven and Ferragut et al. Our results, however, do not agree entirely with some previous reports. McMurtry and Scriven and Ferragut et al. reported that the development duration of *E. hibisci* and *E. stipulatus*, respectively, when fed on *Tetranychus* species, was longer than when fed on citrus red mite. Our data show that all three *Euseius* species developed more rapidly when fed *Tetranychus pacificus* than when fed citrus red mite. With Pacific spider mite as the basic food, the duration of immature development for *Euseius tularensis*, *E. stipulatus*, and *E. hibisci* was 11.4, 17.2 and 16.0 h shorter, respectively, than with citrus red mite as the basic food (Figure 1). The reason probably was that Pacific spider mite used in these tests was primarily in the egg stage. As these eggs were washed from glasshouse-grown lima-bean plants, there was little webbing to impede the movement of the phytoseiid mites. Moreover, some eggs hatched soon after they were placed into test arenas, resulting in prey eggs and larvae, both of which were favored foods for the phytoseiids. On the other hand, the citrus red mite used in the tests was mainly in the egg stage, with few larvae. Previous observations indicated that, with citrus red mite as prey, *E. hibisci* females captured and consumed mainly larvae and small nymphs, with very little feeding on eggs.

Table1. Development duration (h) of three *Euseius* species on different foods and food combinations at 25 ℃

Food provided[1]	Egg		Larva		Nymph (proto-+deuto-)		Immature (all stages)
	Mean	SD[2]	Mean	SD	Mean	SD	Mean
Euseius tularensis							
P	61.2	8.4	26.7	7.4	50.1	15.2	137.6 abc[3]
P+AH	57.6	8.1	25.5	7.8	45.3	11.1	131.4 ab
P+WH	57.6	8.1	24.8	9.1	42.2	12.2	128.5 a
PM	61.2	8.4	30.4	7.6	57.3	8.4	148.6 cd
PM+AH	64.8	8.9	26.9	7.2	56.6	6.1	144.6 bc
PM+WH	64.8	8.9	23.9	5.8	58.9	7.1	144.8 be
RM	59.3	8.7	27.3	10.2	71.8	11.4	160.3 d
RM+AH	61.2	10.1	27.5	8.5	73.2	9.9	158.7 d
RM+WH	57.6	5.8	26.2	8.2	65.4	11.2	153.2 cd
Euseius stipulatus							
P	50.4	8.1	20.8	7.7	45.6	7.7	114.8 a
P+AH	50.4	9.9	19.3	6.0	48.0	10.5	118.7 a
P+WH	49.2	8.4	19.6	7.6	46.4	9.1	118.0 a
PM	57.6	11.4	30.0	5.7	54.1	12.6	140.7 b
PM+AH	56.4	7.6	26.4	5.9	55.6	11.8	138.0 bc
PM+WH	56.4	7.6	29.2	6.6	55.1	11.7	137.1 bcd

续表

Food provided[1]	Egg		Larva		Nymph (proto-+deuto-)		Immature (all stages)
	Mean	SD[2]	Mean	SD	Mean	SD	Mean
RM	52.8	6.8	30.8	4.8	78.8	11.6	158.0 cd
RM+AH	54.0	8.0	26.8	4.7	77.9	9.2	155.1 d
RM+WH	56.4	9.5	26.4	5.5	78.2	11.2	154.2 d
Euseius hibisci							
P	46.8	6.2	18.4	8.5	55.2	6.9	120.0 a
P+AH	48.0	6.3	20.7	12.8	51.7	10.4	118.6 a
P+WH	48.0	6.3	19.9	7.4	55.8	8.8	120.0 a
PM	46.8	10.1	33.0	13.7	63.9	13.5	141.8 bcd
PM+AH	44.4	9.5	29.6	11.9	58.0	13.6	132.0 ab
PM+WH	43.2	10.5	25.6	9.2	60.7	15.1	133.7 abc
RM	49.2	8.4	29.2	9.3	78.1	12.7	154.9 d
RM+AH	48.0	8.5	29.9	4.5	72.9	11.3	150.0 cd
RM+WH	51.6	11.0	28.4	5.5	75.8	13.9	150.6 cd

1 P, pollen; PM, Pacific mite; RM, citrus red mite; AH, aphid honeydew; WH, whitefly honeydew.
2 Standard deviation of mean.
3 Data for each *Euseius* species followed by the same letter are not significantly different ($P=0.05$; Duncan's Test).

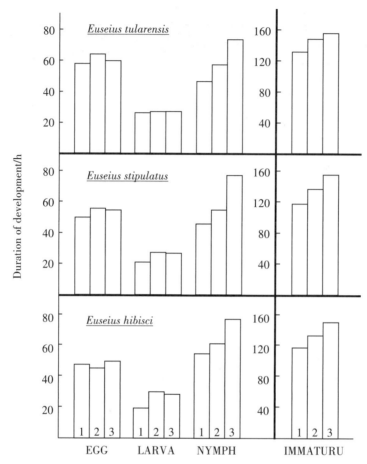

Figure 1. Duration of development (h) of three *Euseius* species when fed on different basic foods (1) with pollen as basic food; (2) with Pacific mite as basic food; and (3) with citrus red mite as basic food. Values represent means of the three treatments (two kinds of honeydew and without honeydew) for each basic food.

Mortality of the three *Euseius* species on different foods is shown in Table 2. With pollen as the basic food, the mortality was low for all species and supplementary foods, while with citrus red mite as food, the mortality was 53.3% and 40.0% for *E. tularensis* and *E. stipulatus*, respectively. Mortality of *E. tularensis* on Pacific mite was higher than that of the other two species on the same prey.

Immature mortality of *E. tularensis* and *E. stipulatus* in the presence of supplementary foods was generally lower than in the absence of supplementary foods, although there were no significant differences ($P=0.05$) between the presence and absence of supplementary foods.

The mortality of the three *Euseius* species mainly occurred in the nymph stages. In the course of the experiments, some of the larvae which did not feed (evidenced by lack of coloration in the gut) molted to the nymphal stage. These nymphs, however, died shortly after molting. Some individuals of certain phytoseiid mite species may molt to the protonymphal stage without feeding, but none reached the deutonymphal stage. Our observations confirmed their results. They also explained why the mortality of the nymphal stage was higher than that of the larval stage.

Table2. Mortality of three *Euseius* species on different foods and food combinations at 25 ℃

Food Provided[1]	Larva		Nymph		Immature
	N[2]	Mortality/%	N	Mortality/%	Mortality/%
Euseius tularensis					
P	30	0	30	3.3	3.3 a[3]
P+AH	31	0	31	0	0
P+WH	30	0	30	3.3	3.3 a
PM	28	0	28	39.3	39.3 bc
PM+AH	29	0	29	24.1	24.1 b
PM+WH	30	3.3	29	20.1	23.3 b
RM	30	20.0	24	41.7	53.3 c
RM+AH	30	20.0	24	25.0	40.0 bc
RM+WH	30	30.0	21	9.5	36.7 bc
Euseius stipulatus					
P	31	3.2	30	0	3.2 a
P+AH	30	6.7	28	0	6.7 a
P+WH	30	0	30	0	0
PM	30	0	30	10.0	10.0 a

续表

Food Provided[1]	Larva		Nymph		Immature
	N[2]	Mortality/%	N	Mortality/%	Mortality/%
PM+AH	30	0	30	13.3	13.3 a
PM+WH	30	0	30	6.7	6.7 a
RM	30	0	30	40.0	40.0 b
RM+AH	30	3.3	29	34.5	36.7 b
RM+WH	30	0	30	33.3	33.3 b
Euseius hibisci					
P	31	3.2	30	0	3.2 a
P+AH	29	0	29	0	0
P+WH	30	3.3	29	0	3.3 a
PM	30	0	30	6.7	6.7 a
PM+AH	30	0	30	0	0
PM+WH	30	0	30	6.7	6.7 a
RM	30	3.3	29	24.1	26.7 b
RM+AH	30	6.7	28	32.1	36.7 b
RM+WH	30	0	30	30.1	30.0 b

1 P, pollen; PM, Pacific mite; RM, citrusred mite; AH, aphid honeydew; WH, whitefly honeydew.
2 Number of individuals in the tests.
3 Data for each *Euseius* species followed by the same letter are not significantly different ($P=0.05$; χ^2 contingency test for % survival).

Oviposition rates (the number of eggs laid per female per day) of all three *Euseius* species in the presence of supplementary foods were higher than in the absence of such foods. Most of the differences were statistically significant between the presence and absence of supplementary foods (Table 3). However, no significant differences occurred between the two types of supplementary foods (honeydew of aphids and whiteflies). McMurtry and Scriven found that the combination of honeydew and either abundant mite prey or pollen stimulated a higher oviposition rate of *E. hibisci* than mite prey or pollen alone. Our results were similar, although there were some differences in the oviposition rates because of the different experimental conditions. Ragusa and Swirski showed that the addition of honeydew increased the oviposition rate of *Amblyseius swirskii* Athias-Henriot on *Tetranychus cinnabarinus* (Boisduval).

The oviposition rates of the three *Euseius* species did not differ significantly between pollen and Pacific mite as basic food, but significant differences did occur between citrus red mite and the other two basic foods. The oviposition rates of all three species, with citrus red mite as the basic food, were lower than pollen or Pacific mite as the basic food. The reason for this probably was because the citrus red mite in the tests was mostly in the egg stage, eggs being unfavorable for these phytoseiids. Tanigoshi et al. found that *Euseius hibisci* did not lay eggs when fed exclusively on citrus red mite. Our observations showed that adult females of the three *Euseius* species, confined on leaves with few larvae and many eggs of citrus red mite, oviposited but ceased reproduction after 5~6 days.

Table 3. Oviposition rate of three *Euseius* species on different foods and food combinations at 25 ℃[1]

Food provided[2]	*Euseius tularensis*		*Euseius stipulatus*		*Euseius hibisci*	
	Mean	SD	Mean	SD	Mean	SD
P	1.22 ab[3]	0.23	1.25 ae	0.29	1.80 a	0.31
P+AH	1.47 ac	0.25	2.13 d	0.17	2.73 b	0.37
P+WH	1.68 c	0.23	1.90 dc	0.21	2.86 b	8.36
PM	1.05 b	0.24	1.48 ab	0.10	1.83 a	0.14
PM+AH	1.52 ac	0.07	1.73 bc	0.03	2.53 b	0.32
PM+WH	1.47 ac	0.11	1.68 bc	0.09	2.78 b	0.20
RM	0.61 d	0.13	0.78 f	0.13	0.64 c	0.05
RM+AH	0.97 b	0.13	1.11 ef	0.17	0.89 c	0.17
RM+WH	1.03 b	0.25	1.17 ae	0.22	0.97 c	0.21

1 With pollen and Pacific mite as the basic foods, the oviposition rate is the number of eggs laid per female per day for a 10-day period; with citrus red mite as the basic food, for a 6-day period.
2 P, pollen; PM, Pacific mite; RM, citrus red mite; AH, aphid honeydew; WH, whitefly honeydew.
3 Data in each column followed by the same letter are not significantly different ($P=0.05$; Duncan's test).

Oviposition rates showed some differences between species on the same food or food combination. The oviposition rate of *E. tularensis* generally was lower than that of the other two *Euseius* species. When pollen served as the basic food, the oviposition rates of *E. tularensis*, *E. stipulatus* and *E. hibisci* were 1.50, 1.76 and 2.46 eggs female^{-1} day^{-1} for a 10-day period, respectively; when Pacific mite served as the basic food, oviposition rates were 1.40, 1.63 and 2.38 eggs respectively; when citrus red mite served as the basic food, oviposition rates were respectively 0.87, 1.02 and 0.82 eggs female^{-1} day^{-1} for a 6-day period.

Honeydew is a commonly available supplemental food in southern California citrus orchards. Sources of this honeydew are primarily aphids, mainly in the spring and early summer, and whiteflies, essentially throughout the year. Other Homoptera, such as lecaniine (soft) scale insects and mealybugs, are probably of minor importance on citrus in most areas, although mealybugs generally are the most common honeydew-producing insects on avocado. Honeydew may play a significant role in the dynamics of *Euseius* populations on citrus, by promoting a higher rate of increase of these predators when spider mites are the main basic food present. Moreover, utilization of these supplemental foods may prolong survival and thus prevent severe declines in the phytoseiid populations during shortages of primary foods.

The results of this study also suggest that laboratory tests to determine reproductive rates of 'generalist' phytoseiids confined with one kind of food (e.g., spider mites) may underestimate their reproductive potential in the presence of that food in the field, as these predators probably supplement a diet of spider mites or thrips with carbohydrate-rich foods that elevate the reproductive rate on the basic foods. Thus, it may be profitable for a phytoseiid to capture and consume a moderately or marginally favorable prey (in terms of reproduction on that food alone) if supplemental foods are available.

Original article was published in *Experimental & Applied Acarology*, volume 8, 1990

Toxicological and Biochemical Characterizations of GSTs in *Liposcelis bostrychophila* Badonnel (Psocop., Liposcelididae)

Dou Wei, Wang Jinjun, Zhao Zhimo

Abstract: The toxicological and biochemical characteristics of glutathione S-transferases (GSTs) in the resistant and susceptible strains of *Liposcelis bostrychophila* were investigated. The two resistant strains were the dichlorvos-resistant strain (DDVP-R) and PH_3-resistant strain (PH_3-R), and the resistance factors were 22.36 and 4.51, respectively. Compared with their susceptible counterparts, the activities per insect and specific activities of GSTs in DDVP-R and PH_3-R were significantly higher. The apparent Michaelis – Menten constant values (K_m) for 1-chloro-2, 4- dinitrobenzene (CDNB) were obviously lower in DDVP-R and PH_3-R (i.e. lower K_m values, 1.562 5 mM[①] for DDVP-R and 0.623 0 mM for PH_3-R) when compared with their susceptible counterpart (K_m=3.552 0), indicating a higher affinity to the substrate CDNB in resistant strains. In contrast, the catalytic activity of GSTs towards CDNB in the susceptible strain was significantly higher than those in resistant strains. It was noticeable that when reduced glutathione (GSH) was used as substrate, GSTs from resistant strains both indicated a significantly declined affinity. For the catalytic activity of GSTs towards GSH, only the V_{max} value in DDVP-R increased significantly compared with that from the susceptible strain, suggesting an overexpression of GST in this resistant strain. The inhibition kinetics of insecticides to GSTs in vitro revealed that dichlorvos and paraoxon possessed excellent inhibition effects on GSTs. The susceptible strain showed higher sensitivity (I_{50}= 0.900 4 mM) to dichlorvos than DDVP-R and PH_3-R (higher I_{50}s, 8.095 5 mM for DDVP-R and 9.334 6 mM for PH_3-R). As for paraoxon, there was a similar situation. The resistant strains both suggested a higher I_{50} (1.873 5 mM for DDVP-R, and 0.429 1 mM for PH_3-R) compared with the susceptible strain (0.294 3 mM). These suggested that an elevated detoxification ability of GSTs developed in the resistant strains.

Key words: *Liposcelis bostrychophila*; GSTs; resistance

1 Introduction

Liposcelis bostrychophila Badonnel, an important stored-product insect pest,

---

① 1 M=1 mol/L。后同。——编辑注

is cosmopolitan in distribution and commonly found in various processed and unprocessed dry foods in households, granaries and warehouses. Outbreaks of *L. bostrychophila* have been reported in humid tropical countries such as Indonesia, Malaysia, Singapore, the Philippines, Thailand, China and India. Routine fumigations of warehouses and storage facilities with methyl bromide have failed to control the pest. In addition, the rapid development of resistance to chemical and physical treatments by the psocid has also been reported.

Glutathione S-transferases (GSTs; E.C. 2.5.1.18) are a major family of detoxification enzymes found in most organisms. GSTs catalyse the conjugation of the thiol group of reduced glutathione (GSH) to the electrophilic centre of lipophilic compounds, generally making the resultant products more water soluble and excretable than the non-GSH conjugated substrates. Meanwhile, they have GSH-dependent peroxidase activity and play a role in intracellular transportation. Generally speaking, GSTs of insects have been divided into two classes (class I, also referred to as the delta class and class II as sigma class) based on their amino acid sequence homology (40% identity to other members within the class) and immunological properties. Very recently, another insect GST class was established (class III), which is also named epsilon-class GST.

Interest in insect GSTs is focused on the role of these enzymes in insecticide resistance. Elevated GST activity has been detected in strains of insects resistant to organophosphates and organochlorines and this enzyme family has recently been implicated in resistance to pyrethroid insecticides. The conjugation of glutathione to organophosphate insecticides results in their detoxification via two distinct pathways. In O-dealkylation, glutathione is conjugated with the alkyl portion of the insecticide, e.g. the demethylation of the tetrachlorvinphos in resistant houseflies. In another mechanism, O-dearylation, glutathione reacts with the leaving group, e.g. the detoxification of parathion and methyl parathion in the diamondback moth *Plutella xylostella* (Linnaeus). The elevation of GST activity in the resistant insect involves upregulation of multiple enzymes belonging to one or more GST classes, or more rarely upregulation of a single enzyme.

As little is known about the mechanisms of insecticide resistance in psocids, information on biochemistry of *Liposcelis* GSTs may assist in formulating strategies in the control of these rapidly proliferating pests. The objectives of this study were to compare biochemical characterizations of GSTs between the susceptible and resistant strains.

2 Materials and methods

2.1 Test insects

Stock colonies of *Liposcelis bostrychophila* were developed from nymphs collected from a wheat warehouse in Chongqing, China in 1990. The insects were reared on an artificial diet consisting of whole-wheat flour, skimmed milk and yeast powder (10∶1∶1) in an air-conditioned room at (27±1) ℃, relative humidity (RH) 75%—80% and a scotoperiod of 24 h. Three strains were maintained. The susceptible strain was not exposed to blended gas or insecticides.

PH_3-R and DDVP-R strains were treated with appropriate PH_3 and DDVP (diluted with acetone), respectively. PH_3-R and DDVP-R were treated 85 times with the treatments carried out at an 1-month interval under the pressure of 75% adult mortality. The resistance factors of PH_3-R and DDVP-R amounted to 4.51 and 22.36 based on the bioassay, respectively. All experiments were conducted under the conditions described above with 3- to 5-day-old adult females.

2.2 Chemicals and insecticides

Reduced glutathione (GSH; Sigma, St. Louis, MO, USA), 1-chloro-2,4-dinitrobenzene (CDNB; Shanghai Chem. Ltd, Shanghai, China), and other biochemical reagents were of regent grades or better; 56% AlP was obtained from Chemical Industry Inc. (Jining, Shandong Province, China). PH_3 hydrolysed from AlP was the active ingredient. Insecticides used were 80% dichlorvos (DDVP; Pesticide Co. Ltd, Shalong, Hubei Province, China), 98.5% paraoxon (Chem.

Service, West Chester, PA), 5% cypermethrin (Chem. Co. Ltd, Shenglian, Shanghai, China) and 48% chlorpyrifos (Dow AgroSciences LLC, Indianapolis, IN, USA).

2.3 Enzyme preparation

Thirty adults were homogenized manually in 300 μL Tris-HCl buffer (0.05 M, pH=7.5) and centrifuged for 5 min at 4 ℃ and 10 000 g. The resulting supernatants were used as the enzyme sources.

2.4 GST activities

Glutathione S-transferase activities were determined using CDNB and reduced GSH as substrates according to Habig et al. with slight modifications. The total reaction volume per well of a 96-well microtitre plate was 300 μL, consisting of 100 μL of each supernatant (10 psocid equivalents), CDNB [1% ethanol (v/v) included] and GSH in buffer, giving final concentrations of 0.2 and 2 mM of CDNB and GSH, respectively. The non-enzymatic reaction of CDNB with GSH measured without homogenate served as control. The change in absorbance was measured continuously for 5 min at 340 nm and 37 ℃ in a Thermomax kinetic microplate reader (Tecan Austria GmbH, Grödig, Austria). Changes in absorbance per minute were converted into nmol CDNB conjugated/min/mg protein using the extinction coefficient of the resulting 2, 4-dinitrophenyl-glutathione: ε_{340nm}= 9.6 mM/cm.

2.5 Kinetics of GSTs

Values of K_m and V_{max} of total GST from three strains of *Liposcelis bostrychophila* (susceptible, PH_3-R and DDVP-R) were determined for CDNB and GSH, respectively. The activity was recorded with a range of concentrations of CDNB (or GSH) while GSH (or CDNB) was kept constant at 2 mM (or 0.2 mM). K_m and V_{max} values were calculated by with SPSS 10.0 for Windows using the Michaelis – Menten equation.

2.6 Protein content assay

Protein contents of the enzyme homogenate were determined according to the method of Bradford using bovine serum albumin as the standard. The measurement was performed with the wavelength of 340 nm.

2.7 In vitro inhibition of GSTs

To verify whether the evaluated levels of GSTs activity, present in DDVP-R and PH_3-R, plays a role in resistance to insecticides, four chemicals (organophosphates: dichlorvos and paraoxon; pyrethroids: cypermethrin; carbamates: chlorpyrifos) were selected to assess in vitro inhibition of GSTs. Mass homogenates from *L. bostrychophila* were used to study the inhibition of GST activity by different compounds. Stock solutions of the inhibitors were prepared in ethanol and diluted with Tris-HCl (50 mM, pH=7.5), thus the highest ethanol concentration was 1% in the test solutions. One hundred μL of the enzyme source (10 psocid equivalents) and 50 μL inhibitor solution with appropriate concentration range were incubated for 10 min at 30 ℃ and then added to the substrate mixture as described above.

3 Results

3.1 Assay of GST activities and kinetic parameters

Compared with the susceptible strain, the amounts of protein per individual from the two resistant strains both decreased significantly (table 1). But there was no significant difference of individual protein amount between the two resistant strains.

The kinetic parameters of GSTs from *L. bostrychophila* are given in table 2. For both the resistant strains, GSTs showed a significantly higher affinity (i.e. lower K_m values, 1.562 5 mM for DDVP-R and 0.623 0 mM for PH_3-R) to the substrate CDNB than that of susceptible strain (K_m=3.552 0) ($P<0.05$). In contrast, the catalytic activity of GSTs towards CDNB in the susceptible strain was higher (i.e. higher V_{max} value, 0.754 4 mmol/min) than those in the two resistant strains (0.625 5 for DDVP-R, and 0.503 0 for PH_3-R). Despite the significance between PH_3-R and susceptible strain ($P<0.05$), there was no significant distinction between DDVP-R and the other two strains. It was noticeable that when GSH was used as the substrate, GSTs from the two resistant strains both indicated a declined affinity ($P<0.05$). The K_m values for DDVP-R and PH_3-R reached 62.260 1 and 19.435 1 mM, respectively. For the catalytic activity of GSTs towards GSH, only the V_{max} value in DDVP-R increased significantly compared with that from the susceptible strain ($P<0.05$). The activities per insect and specific activities of GSTs in the two resistant strains were significantly higher than their susceptible counterparts ($P<0.05$) (table 1). Between the two resistant strains, the higher activity per insect and higher specific activity of GSTs in DDVP-R were observed (table 1).

Table1. Comparison of the GSTs activity of *L. bostrychophila*

Strain	Protein content (μg/psocid)	Activity of GSTs (nmol/min)	Specific activity of GSTs (nmol/min/mg pro)
Susceptible	7.630 6±0.162 9 a	0.356 0±0.002 8 a	47.593 6±1.170 9 a
DDVP-R	5.014 6±0.091 4 b	0.496 4±0.010 4 b	100.923 4±3.263 8 b
PH_3-R	5.669 7±0.092 3 b	0.400 9±0.014 5 c	71.897 3±3.247 7 c

Each value represents the mean (±SE) of 48 replications.
Mean values within the same column followed by different letters are significantly different ($P<0.05$).

Table2. The K_m and V_{max} values of GSTs from *L. bostrychophila*

Strains	GSH		CDNB	
	K_m(mM)	V_{max}(mmol/min)	K_m(mM)	V_{max}(mmol/min)
Susceptible	8.181 8±0.577 4 a	0.909 1±0.048 2 a	3.552 0±0.436 2 a	0.754 4±0.065 7 a
DDVP-R	62.260 1±2.986 9 b	3.448 3±0.220 3 b	1.562 5±0.115 4 b	0.625 5±0.118 1 ab

续表

Strains	GSH		CDNB	
	K_m(mM)	V_{max}(mmol/min)	K_m(mM)	V_{max}(mmol/min)
PH$_3$-R	19.435 1± 1.154 7 c	0.854 7±0.073 1 a	0.623 0±0.024 2 b	0.503 0±0.030 8 bc

Each value represents the mean (±SE) of five replications.
Mean values within the same column followed by different letters are significantly different ($P<0.05$).

3.2 In vitro inhibition of GSTs

The inhibition kinetics of dichlorvos and paraoxon to GST activities are shown in figure 1 and figure 2, respectively. Among the three strains, the percentage inhibitions of enzyme activity were between 6% and 82%. The efficiencies of the tested inhibitors were compared based on their I_{50}s (the concentration required to inhibit 50% of GSTs activity). The susceptible strain showed higher sensitivity (I_{50}=0.900 4 mM) to dichlorvos than DDVP-R and PH$_3$-R (higher I_{50}, 8.095 5 mM for DDVP-R and 9.334 6 mM for PH$_3$-R). As for paraoxon, there was a similar situation. The resistant strains both suggested a higher I_{50} (1.873 5 mM for DDVP-R, and 0.429 1 mM for PH$_3$-R) compared with the susceptible strain (0.294 3 mM).

The effects of cypermethrin and chlorpyrifos on GSTs in the susceptible and resistant strains are compared in figure 3 and figure 4, respectively. As for cypermethrin, with the increase of concentration, the activities in various strains all declined first and then ascended ($P<0.05$). Under lower concentration ($<0.02\mu M$), however, there was no significant effect difference on GST activities among the three strains (Figure 3). Compared with the resistant strains, cypermethrin possessed a significantly larger inhibition effect on the susceptible strain at a concentration of 0.200 2 μM ($P<0.05$), with no significant effect difference between DDVP-R and PH$_3$-R.

For chlorpyrifos, no obvious inhibition effect was observed (Figure 4). It was noticeable that there was some facilitated effect of chlorpyrifos with a lower concentration on GSTs activities. ANOVA showed that the effect was significantly higher to GSTs in the susceptible strain than those from DDVP-R and PH$_3$-R at 0.002 3 and 0.022 8 μM ($P<0.05$).

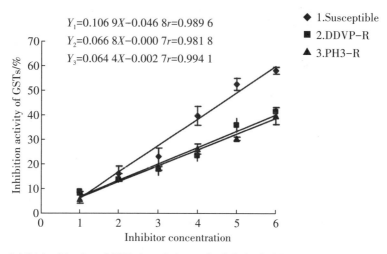

Figure1. Inhibition kinetics of GSTs from *L. bostrychophila* by dichlorvos. Note: 1, 2, 3, 4, 5, 6 corresponding to 0.006, 0.030 2, 0.060 3, 0.301 7, 0.603 3 and 3.016 6 mM, respectively. Each point represents the mean of five determinations ($n=5$). Error bars represent the mean±SD

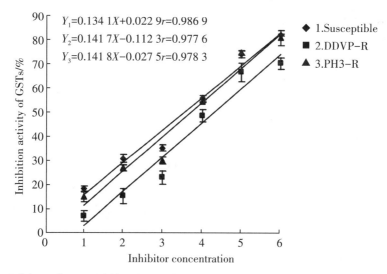

Figure2. Inhibition kinetics of GSTs from *L. bostrychophila* by paraoxon. Note: 1,2,3,4,5,6 corresponding to 0.000 8, 0.007 7, 0.077 1, 0.771 0, 7.709 5 and 77.095 3 mM, respectively. Each point represents the mean of five determinations ($n=5$). Error bars represent the mean±SD

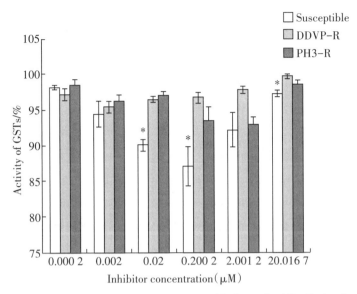

Figure 3. The effects of cypermethrin on GSTs activity of *L. bostrychophila*. Each value represents the mean of five determinations ($n=5$). Error bars represent the mean±SD. Within the same concentration, bars of the susceptible strain with asterisk are significantly different from the resistant strains ($P<0.05$)

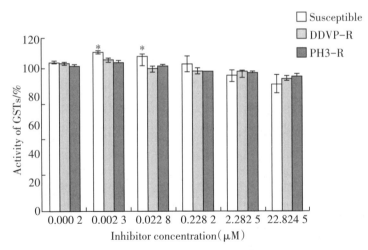

Figure 4. The effects of chlorpyrifos on GSTs activity of *L. bostrychophila*. Each value represents the mean of five determinations ($n=5$). Error bars represent the mean±SD. Within the same concentration, bars of the susceptible strain with asterisk are significantly different from the resistant strains ($P<0.05$)

4 Discussion

Glutathione S-transferases have been described to play a major role as a detoxification mechanism for insecticides, thus contributing to insecticide resistance in economically important pest species in diverse agronomic cropping systems as well as disease vectors such as mosquitoes. They are involved in the O-dealkylation or O-dearylation of organophosphorus insecticides, as a secondary mechanism in the detoxification of organophosphate metabolites and in the dehydrochlorination of organochlorines. Recent studies on recombinant Anopheles class I GST enzymes have also shown that they recognize pyrethroids as either substrates or inhibitors, and there is now evidence that they are directly involved in pyrethroid resistance in the planthopper *Nilaparvata lugens*.

In many cases, the individual GST enzyme(s) involved in resistance have not been identified and GSTs have been implicated by association only (i.e. an increase in GST activity, detected using model substrates, in insecticide-resistant strains of insects vs. their susceptible counterparts). In *Musca domestica*, for instance, a 30-fold difference of GST activities between susceptible and resistant strains was observed. An increased rate of DDT dehydrochlorination in the resistant strain of *A. gambiae* was associated with quantitative increases in multiple GST enzymes and one of these, *GSTe2*, encoded an enzyme that has the highest levels of DDT dehydrochlorinase activity reported for any GST. In the diamondback moth, *Plutella xylostella*, increased expression of the *PxGST3* gene, which encoded an enzyme capable of degrading organophosphorous insecticides, was strongly correlated with resistance. Elevated levels of GST activity were very recently also shown to be associated with spider mite resistance to acaricides, particularly in abamectin-resistant strains of the two-spotted spider mite, *Tetranychus urticae*.

In the present study, the biochemical characterization analysis of GSTs from the susceptible and resistant strains showed that elevated activities and higher specific activities of GSTs were both observed in the two resistant strains. Similar phenomena were reported in many other resistant insects above, such as 0.41 μmol/(min·mg) of protein of GST-specific activity in resistant strain vs. 0.267 μmol/(min·mg) of protein in susceptible strain of *Nilaparvata lugens*. This indicated that in *Liposcelis bostrychophila*, there was some correlation between the development of resistance and the quantity of GSTs.

Apparent K_m values for CDNB were significantly lower for DDVP-R and PH_3-R compared with the susceptible strain. Compared with most invertebrate GSTs investigated, these K_m values for CDNB can be considered as quite high. In contrast, varying GSH concentrations at a constant CDNB concentration revealed there were higher K_m values in DDVP-R and PH_3-R. V_{max} values were approximately four times as high for DDVP-R compared with the susceptible strain, indicating an overexpression of GSTs in this strain. Such an overexpression of GSTs as a consequence of constant insecticide selection pressure was also described in other insect pests, e.g. *Drosophila melanogaster*, *Musca domestica*, *A. gambiae*, and *Aedes aegypti*.

The in vitro inhibition study showed that only dichlorvos and paraoxon possessed excellent inhibition effects on GSTs, and the susceptible strain showed higher sensitivity to dichlorvos and paraoxon than the resistant strains. This further confirmed that constant selection pressure with dichlorvos and PH_3 induced higher GST activity. Inhibitions of CDNB-conjugating activity by organophosphorus malathion and malaoxon were also reported in *A. dirus* and *A. gambiae*, and the inhibition percentage of malathion both amounted to 72.7% while a relatively lower effect of 33.7% was observed by malaoxon. In contrast, the inhibition effect of cypermethrin on GSTs was quite poor, and only at a higher concentration of 0.200 2 μM had some impact, and even so, the inhibition effect was <15%. It was noticeable that for chlorpyrifos, there was some facilitated effect of chlorpyrifos with a lower concentration on GSTs activities, and in *A. dirus* and *A. gambiae*, there was no detectable inhibition effect by carbamates (propoxur and bendiocarb applied). Hence it may be considered that GSTs cannot metabolize chlorpyrifos in vitro.

In general, the present study has provided some basic information on the GSTs of *Liposcelis bostrychophila*. This will contribute to the complete understanding of the mechanisms of insecticide resistance of *L. bostrychophila* in the future. Further studies of *L. bostrychophila* GSTs will be necessary to embark on the purification of enzymes and the subsequent explorations of molecular genetic mechanisms responsible for resistance.

Original article was published in *Journal of Applied Entomology*, volume 130, 2006

Infection by *Wolbachia* Bacteria and Its Influence on the Reproduction of the Stored-Product Psocid, *Liposcelis tricolor*

Dong Peng, Wang Jinjun, Zhao Zhimo

Abstract: *Wolbachia* are maternally inherited intracellular bacteria that infect a wide range of arthropods and nematodes and are associated with various reproductive abnormalities in their hosts. The infection by *Wolbachia* of the psocid, *Liposcelis tricolor* (Pscoptera: Liposcelididae), was investigated using long PCR amplification of the *wsp* gene that codes for a *Wolbachia* surface protein. The results showed that *Liposcelis tricolor* was positive for *Wolbachia*. Phylogenetic analysis showed that the *Wolbachia* found in *Liposcelis tricolor* was related to the B-group. *Wolbachia* infection in *Liposcelis tricolor* could be removed through antibiotic treatment. The results of crosses including $♀^{w+}×♂^{w+}$, $♀^{w-}×♂^{w+}$, $♀^{w+}×♂^{w-}$ and $♀^{w-}×♂^{w-}$, suggested that the removal of *Wolbachia* resulted in lower egg production by *Liposcelis tricolor*. The mean embryonic mortality of offspring produced by *L. tricolor* without *Wolbachia* was significantly higher than that of control.

Key words: endosymbionts; molecular detection; antibiotic treatment

Introduction

Wolbachia, rickettsia-like proteobacteria, are found in the reproductive tissues of a wide range of arthropod species. *Wolbachia* are obligate intracellular parasites that are transmitted maternally from infected females to their progeny. *Wolbachia* infection are associated with a variety of reproductive anomalies in the host, including cytoplasmic incompatibility between different populations or closely related species, thelytoky or parthenogenesis induction in parasitoid wasps, male-killing in coccinellid beetles, and feminization of genetic males in isopods.

Wolbachia cannot be cultured in defined media and detection within infected gonadal cells may be time consuming. Therefore, detection of *Wolbachia* infection has been based largely on amplification of *Wolbachia* DNA using allele-specific polymerase chain reactions. Primers designed from the 16S rDNA, *ftsZ* and *wsp* genes have been used to amplify *Wolbachia* DNA from a diverse array of arthropods.

Antibiotic treatment is one established method for producing *Wolbachia*-free individuals. A wide range of antibiotics with different modes of action have been used to produce aposymbiotic insects. The strains of parthenogenetic *Trichogramma* wasps revert to sexual reproduction when rickettsial endosymbionts are eliminated from wasp gonads.

The psocid, *Liposcelis tricolor*, a bisexual species, is worldwide and commonly found in various processed and unprocessed dry foods in households, granaries, and warehouses. They transmit harmful microorganisms, including fungi and bacteria on their bodies, hair, and in feces. Psocids are found to be harmful pests on seed goods as well by causing damage to seed kernels. This study was initiated to find if *Wolbachia* infest *Liposcelis tricolor* and their influence on the reproductive biology of *L. tricolor*.

Materials and methods

Insects

Stock colonies of *Liposcelis tricolor* originated from larva collected in a wheat warehouse in Shandong, China in 2003. This colony was reared on an artificial diet consisting of whole wheat flour, skim milk and yeast powder (10:1:1) in an air-conditioned room at (27±1) ℃ and a scotoperiod of 24 h. Cultures were set up in glass bottles (250 mL) with a nylon screen cover and kept in desiccators (5 liter), in which the humidity was controlled with saturated NaCl solution at 75%—80% RH. After several generations, insects from the stock colony were used for the tests.

DNA extraction

Twenty adult *L. tricolor* were surface-sterilized in a series of double distilled water and 70% ethanol washes, then frozen under liquid nitrogen and crushed in 300 μL DNA extraction buffer (100 mM Tris-HCl, pH=8.0, 50 mM NaCl, 50 mM EDTA, 1% SDS, 0.15 mM spermine, 0.5 mM spermidine), homogenized using a DNA-free disposable polypropylene pestle and incubated with 2 μL of proteinase-K (20 mg/L) for 2 h at 50 ℃, followed by 5 min at 95 ℃ for denaturing. One volume of phenol saturated water (pH=8.0) and 1 volume of chloroform: isoamyl alcohol (24:1) was added before centrifugation for 10 min at 10 000 rpm[①]. The supernatant was collected and gently mixed with 0.2 volume of Na-acetate (3 mM, pH=5.2) and 2 volumes of 100% ethanol. After precipitation for 2 h at −20 ℃, the DNA was washed with 70% ethanol, air dried and finally resuspended in 20 μL of ddH$_2$O. 3 μL of the DNA sample was used for PCR experiments. DNA quality was determined using insect 12S rDNA primers. The PCR mixtures and cycling conditions were from Yoshizawa and Johnson.

PCR screening procedure for *Wolbachia* infection

Long PCR was performed in a 25 μL containing 50 mM Tris (pH=9.2), 16 mM ammonium sulphate, 1.75 mM MgCl$_2$, 350 μM dNTPs, 800 pM of primers (*wsp*-F, 5'-TGG TCC AAT AAG TGA TGA AAG AAA CTA GCT A and *wsp*-R, 5'-AAA AAT TAA ACG CTA CTC CAG CTT CTG CAC), 1 unit of *Pwo* and 5 units of *Taq* DNA polymerases. The Long PCR was carried out using three linked profiles: (i) one cycle (2 min at 94 ℃), (ii) 10 cycles (10 s at 94 ℃, 30 s at 65 ℃, 1 min at 68 ℃), (iii) 25 cycles (10s at 94 ℃, 30s at 65 ℃, 1 min at 68 ℃, plus an additional 20 s added for every consecutive cycle).

To confirm that the PCR products obtained were not due to contamination, an attempt was made to sequence the long PCR products directly from *Liposcelis tricolor*. The PCR product was then ligated into a pGEM-T vector. Three clones were produced that were sequenced by Shanghai Sangon Biological Engineering Technology & Services Company.

---

① rpm现多写作r/min。——编辑注

Phylogenetic analysis

The *wsp* datasets representative sequences for *Wolbachia* groups in different hosts were retrieved from GenBank and included in phylogeny reconstruction. Sequences were aligned using CLUSTAL W. Phylogenies for *wsp* were estimated by neighbor-joining analysis of the sequence data. Gaps were coded as missing data. Distances were calculated using maximum likelihood. The model of sequence evolution and the parameter values best fitting the data for each locus were identified using likelihood ratio tests, as implemented in PHYLIP 3.57c to search for the tree with the highest likelihood. The parameter values suggested by PHYLIP for each dataset were used to specify the distance matrix for neighbor-joining analysis on 1 000 bootstrap replicates.

Antibiotic experiments

Rifampicin, a potent inhibitor of DNA-dependent RNA polymerase of bacteria, was used to produce *Wolbachia*-free hosts. Rifampicin (1%) was prepared by mixing 10 g of the standard diet with 0.1 g rifampicin (Sigma, www.sigmaaldrich.com). Three grams of this food-antibiotic mixture were further diluted into 10 and 100 g of standard diet to make final concentrations of 1%, 0.3%, and 0.03% antibiotic in the standard diet. Dry mixing of the food and antibiotic powder proved unsatisfactory and instead the food-antibiotic mixture was moistened with sterile distilled water, well stirred to create homogeneous slurry, air-dried and ground to a powder. After 4 weeks treatment with rifampicin, the psocids were checked with the above molecular method. After two generations, the psocids of *Wolbachia*-free strain (W-) and untreated stain (W+) were used for the following cross experiment.

Effects of *Wolbachia* removal on fecundity and fertility

Four crosses ($♀^{W+} × ♂^{W+}$, $♀^{W-} × ♂^{W+}$, $♀^{W+} × ♂^{W-}$, and $♀^{W-} × ♂^{W-}$) were tested in this experiment. The cross of the infected female and male was used as a control. In each cross experiment, 30—61 maturing females (when the colorless stage begins to change to brown, 12—24 h after the final molt) were removed from culture populations with a fine brush. They were placed individually in a small glass vials containing a piece of filter paper on one side of which was glued the culture medium. The glass vials were covered with fine mesh net to prevent psocids escape. One corresponding matured male was placed into each glass vials. The glass vials were placed in each humidified desiccator (75%—80%RH). The desiccators were placed in growth chambers at 27.5 ℃ with a scotoperiod of 24 h. Eggs produced from the different treatment were counted, and the mortality of eggs from the different treatments was monitored. The total egg numbers within 60 days were counted for each experiment, and the statistical analysis was conducted using SPSS statistical package.

Results

Genomic DNA prepared from *Liposcelis tricolor*, was amplified by Long PCR, using the primers (*wsp*-F and *wsp*-R). A 605 bp fragment of the *wsp* gene was amplified (Figure 1) and this sequence was deposited in GenBank with accession number AY639593. The infection frequency of *Wolbachia* in *Liposcelis tricolor* was found to be 100% in 5 randomly selected females and 5 males.

The phylogenetic tree based on the *wsp* sequences using neighbor-joining after bootstrapping 1 000 times. The topology showed the division of *Wolbachia* into two subgroups, A and B. The A subgroup included fourteen species and the B subgroup included fifteen species. *Wolbachia* in *Liposcelis tricolor* belonged to the B subgroup and was close to *Wolbachia Pip* in *Culex quinquefasciatus* and *Aedes albopictus*.

After 4 weeks treatment with 1% rifampicin, no detectable level of the *wsp* gene was amplified, while the same gene fragment remained present in the 0.3% or 0.03% rifampicin treated individuals (Figure 2). A strain of *Liposcelis tricolor* without *Wolbachia* infection was obtained.

In cross experiments, all combinations could produce eggs. Total egg production was significantly increased in crosses with *Wolbachia* present, and egg mortality was significantly lower (Table 1, Figure 3).

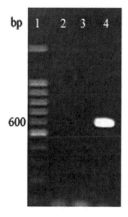

Figure1. Long PCR amplified *wsp* sequences from *Liposcelis tricolor*. Lane 1, 100 bp DNA marker; lane 2 and 3 no DNA-control; lane 4, *L. tricolor*

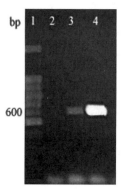

Figure2. Long PCR amplification of *wsp* sequences from *L. tricolor* treated with antibiotic of different dosages. Lane 1, 100 bp DNA marker; Lane 2, 1% rifampicin; lane 3, 0.3% rifampicin; lane 4, 0.03% rifampicin

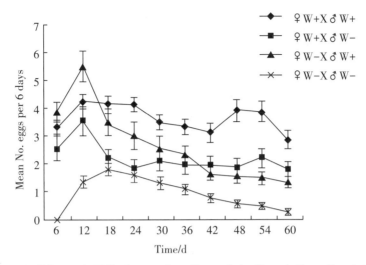

Figure3. Effects of *Wolbachia* on fecundity and fertility of *Liposcelis tricolor*

Discussion

In preliminary experiments both regular and long PCR protocols were used to detect the infection of *Wolbachia* in *Liposcelis tricolor* in the present study, and the target fragment was scored in both protocols. However, there were some non-specific fragments that appeared using the regular PCR protocol. Therefore, the long PCR protocol was used throughout the experiment.

Phylogenetic evidence has shown that horizontal transfer of *Wolbachia* must have occurred in the course of evolution because closely related bacterial strains can be found in unrelated hosts. In this study, phylogenetic analysis showed that *Wolbachia* in *Liposcelis tricolor* was close to *w*Pip present in *Culex quinquefasciatus* and *Aedes albopictus*.

In the present study, a strain of *Liposcelis tricolor* without *Wolbachia* infection was obtained by treating with 1% rifampicin for 4 weeks. Lower concentrations of rifampicin (0.3% or 0.03%) failed to remove the *Wolbachia* infection completely. Similar results were reported for tetracycline treatment of *Wolbachia* infection in *Aedes albopictus* and rifampicin treatement of *Liposcelis bostrychophila* by Yusuf and

Turner.

Wolbachia have a variety of different effects on their hosts' reproduction. Cytoplasmic incompatibility is the most widespread and, perhaps, the most prominent feature that *Wolbachia* endosymbionts impose on their hosts. Cytoplasmic incompatibility resulted in embryonic mortality in crosses between insects with different *Wolbachia* infection status. It can be either unidirectional or bidirectional. Unidirectional cytoplasmic incompatibility is typically expressed when an infected male is mated with an uninfected female. However, all cross combinations in this study could produce hatching eggs, and the cross $♀^{w-} × ♂^{w+}$ did not show higher embryonic mortality compared to the $♀^{w-} × ♂^{w-}$ cross. As a result there is not enough evidence to confirm cytoplasmic incompatibility phenomenon caused by Wolbachia in *Liposcelis tricolor*.

Wolbachia could have positive effects on host fecundity and fertility. In this study, egg production of $♀^{w-}$ mated with $♂^{w-}$, $♀^{w+}$ with $♂^{w-}$ and $♀^{w-}$ with $♂^{w+}$ were all significantly less than the control. The eggs with the highest survival were from the control crosses with both male and female infected, and the other crosses are statistically distinguishable. This suggests that for optimal reproduction, *Wolbachia* must also be present in the male. It implies that *Wolbachia* infection can have positive effects on fecundity and fertility of *Liposcelis tricolor*. Similarly, Dedeine et al. reported that the wasp *Asobara tabida* without *Wolbachia* removed by antibiotics also had lower egg production. By comparison, the mortality of eggs produced by *Liposcelis tricolor* without *Wolbachia* infection was significantly increased. This could be due to cooperative evolution between *Wolbachia* and the psocid host.

In recent years, there has been increasing interest in the biology of *Wolbachia* and in its application as an agent for control or modification of insect population. Therefore, the mode of action of *Wolbachia* manipulation on psocid reproduction warrants further investigation.

Original article was published in *Journal of Insect Science*, volume 6, 2006

论文选集

主题三

螨类毒理学研究

桔全爪螨(*Panonychus citri* McGregor)实验种群生命表的组建与分析

赵志模,朱文炳,叶辉,李强

摘要:本文通过在三种恒温条件下对桔全爪螨进行饲养,分别组建其实验种群生命表,并应用 Morris 模式和 Leslie 矩阵,以探讨温度与桔全爪螨种群数量变动的关系;组建 25 ℃下的实验种群生殖力表,以计算桔全爪螨的内禀增长能力、周限增长率和净生殖率等指标。

Construction and Analysis on Life Table for Experimental Population of *Panonychus citri* McGregor

Zhao Zhimo, Zhu Wenbing, Ye Hui, Li Qiang

Abstract: This paper deals with the life table of *Panonychus citri* McGregor at three constant temperatures of 20, 25 and 30 ℃ and analysis the relationship between temperature and population dynamics of the mite with Morris mathematical model and Leslie matrix model. The results of the experiments show that the influence of temperature on the population dynamics of the mite is in three aspects: 1, the survival percentage of whole generation of the mite; 2, the sex ratio; 3, the average numbers of the eggs laid by the females. The last two are the most important to the index of population trend "I". The simulation tests on computer by Leslie matrix model show that the different characteristics of the number and constitution of the mite population varies with the sequence of time.

应用生命表技术是研究生物种群数量动态的一个重要手段。迄今为止,国内关于昆虫种群生命表的研究已有不少报道,但对螨类种群生命表的研究却甚为罕见。本文通过在 20、25 和 30 ℃三种恒温条件下室内饲养桔全爪螨,分别组建其实验种群生命表,以探讨温度与桔全爪螨种群数量变动的关系;通过适当的数据处理,把 Morris 模式和 Leslie 矩阵结合起来用于生命表数据分析,以阐明与该螨种群结构及年龄组配有关的生殖力和死亡率之间的关系;组建25 ℃下实验种群生殖力表,计算桔全爪螨的内禀增长能力、周限增长率和净生殖率等指标。

材料及方法

(1)供试材料:三月下旬在本院桔园采集桔全爪螨若螨或成螨,在拟处理温度下室内集群饲养,以雌成螨产下的卵作为初始供试材料。

(2)饲养皿的设置:在直径为9 cm的表皿内,置一层浸水的塑料泡膜,上面平铺一张滤纸,其上放一片平展的甜橙叶,叶四周用浸过水的脱脂棉条包围,将供试螨挑在叶片上饲养,叶片每三天更换一次。

(3)试验处理:选用PYX-DHS恒温箱,设置20 ℃、25 ℃和30 ℃(±0.5 ℃)三种温度,将饲养皿放在恒温箱内进行饲养实验,每天给予12 h光照(约600 lx)。

(4)饲养方法和观察记载:将雌成螨在2 h内新产下的卵,放入饲养皿内,集群饲养于恒温培养箱中。各温度供试卵数:20 ℃下29粒,25 ℃下62粒,30 ℃下29粒。从进入实验之时起,每8 h观察一次,记载孵化卵数,并将初孵幼螨用细毛笔挑入新的培养皿内,每皿一螨,置于相应处理温度下,仍然每8 h观察一次,记载幼螨、若螨的死亡数量,一待进入成螨阶段,则进行雌雄配对,每8 h观察记载一次雌成螨的产卵数,并将卵挑出,直到雌螨死亡为止。

结果及分析

(1)生命表的组建及分析

实验种群生命表是以种群的虫态或龄期划分时间间隔的,以各发育阶段的死亡率、繁殖阶段的繁殖数量为主要内容组建而成的。桔全爪螨的发育阶段用x表示,可详细区分为卵、幼螨、一伏期、前若螨、二伏期、后若螨、三伏期及成螨等八个阶段,但鉴于一伏期至三伏期五个阶段不便区分,同时其中有的阶段历期甚短,因此在组建生命表时,只区分为卵、幼螨、若螨、成螨四个发育阶段。进入x期的存活螨数用l_x表示,在x期内死亡的螨数用

d_x表示,其死亡率用$100q_x$表示,显然:

$$l_{x+1}=l_x-d_x,\ 100q_x=\frac{d_x}{l_x}\times 100$$

其中成螨期的d_x,是针对雄螨对产卵量没有直接关系而作为成螨的一个死亡因子设置的。但是在计算上不是把雄性占的比例作为死亡率直接扣除,而是假设在理想的情况下,性比为1∶1,这样成螨就没有性比这个死亡因子造成的死亡数。如果性比不是1∶1,则在成螨下方设置"雌螨×2"栏,它的l_x用下式求出:

"雌螨×2"的l_x=成螨的l_x×雌性比×2,这样成螨的d_x就可以由"成螨"的l_x减去"雌螨×2"的l_x得出。显然成螨的d_x有三种可能的值:

当性比为1∶1时,d_x=0;

当雌性比小于50%时,d_x为正值表示死亡数量;

当雌性比大于50%时,d_x为负值,表示应多存活的数量。

试验结果表明,在不同温度下,桔全爪螨的性比有明显差异,在生命表中,成螨的d_x就恰好提供了这个信息。

为了反映不同温度下雌成螨在产卵量上的差异,生命表中还设置了"正常雌螨×2"这个栏目。所谓"正常雌螨"是指能达到标准产卵量的雌螨。本生命表,假设标准产量为60粒。显然,由于温度等因子的影响,不少雌螨是达不到标准产卵量的,根据该代每个雌螨实际平均产卵量同标准产卵量的差额,就可以折算出"不正常雌螨"的数量,作为死亡数扣除,剩下的便是"正常雌螨"的数量。在计算方法上:"正常雌螨×2"的l_x="雌螨×2"的l_x×$\frac{每雌平均产卵量}{标准产卵量}$。

对于尚未产卵就死亡的雌螨,这相当于实际产卵量为零,这个信息已经反映在每雌平均产卵量中,因此在生命表中不再设置新的栏目。

按照以上原则和计算方法组建的桔全爪螨实验种群生命表如表1。

表1　不同温度下桔全爪螨实验种群生命表

发育阶段x	温度								
	20 ℃			25 ℃			30 ℃		
	l_x	d_x	$100q_x$	l_x	d_x	$100q_x$	l_x	d_x	$100q_x$
卵	29	4	13.8	62	4	6.5	29	5	17.2
幼螨	25	4	16.0	58	6	10.3	24	4	16.7
若螨	21	7	33.3	52	13	25.0	20	4	20.0
成螨	14	−2	−14.3	39	−15	−38.5	16	−8	−50.0
雌螨×2	16	12.6	78.8	54	32.3	59.8	24	16.0	66.7
正常雌螨×2	3.4			21.7			8.0		

说明		20 ℃	25 ℃	30 ℃
	雌性比	57.1%	69.2%	75.0%
	平均产卵量	12.75	24.10	20.00
	标准产卵量		60	

由表1提供的信息,可以用以下方法计算出几个影响桔全爪螨种群动态的主要指标(表2)。

下代期望卵量="正常雌螨×2"的l_x×标准卵量/2

种群趋势指数=下代期望卵量/上代初始卵量

世代总存活率="成螨"的l_x/上代初始卵量

表2　桔全爪螨种群动态的几个主要指标

项目	温度		
	20 ℃	25 ℃	30 ℃
上代初始卵量	29	62	29
下代期望卵量	102	651	240
种群趋势指数	3.52	10.50	8.28
世代总存活率	0.48	0.63	0.55

综合分析表1和表2,可以看出不同温度对桔全爪螨种群动态的影响。其中种群趋势指数(I)是直接反映种群动态的一个重要指标。I值的基本

含义是次代种群数量为当代种群数量的倍数。显然,当$I=1$时,相继两个世代的种群数量相同,表明种群死亡和繁殖维持着动态平衡;当$I>1$时,次代种群数量上升;当$I<1$时,次代种群数量下降。表2的数据说明,如果排除天敌、雨水等其他环境条件的影响,桔全爪螨在20 ℃、25 ℃和30 ℃三种恒温条件下,次代种群数量都呈增长的趋势,其中以25 ℃条件下增长倍数最高,30 ℃时次之,20 ℃时增长缓慢。

在不同恒温条件下,种群趋势指数的差异主要通过三个因子起作用,一是桔全爪螨世代总存活率的高低,二是平均每头雌螨产卵量的大小,三是雌性比例上的不同。从表1和表2明显看出,除了在30 ℃下雌性比较高外,其余各项因子都以25 ℃下对种群的增长更为有利。

Morris(1963)根据特定年龄生命表提供的信息,认为种群趋势指数(I)是各虫态或发育阶段存活率(S_i)、平均每雌产卵量($F \cdot P_F$)和雌性比($P_♀$)的函数,并用以下数学模式来表达:

$$I = \sum_{i=1}^{n} S_i \times F \times P_F \times P_♀$$

式中S_i表示各发育阶段的存活率即$(1-d_x)$;F为标准产卵量;P_F为达到标准产卵量的百分率。

显然,应用Morris模式所计算的结果和表2计算的数据是完全一致的。(如表3)

表3　不同温度下桔全爪螨各发育阶段存活率及I值

温度	组分						
	S_1/%	S_2/%	S_3/%	$P_♀$/%	P_F/%	F	I值
20 ℃	0.862	0.840	0.667	0.571	0.213	60	3.52
25 ℃	0.935	0.897	0.750	0.692	0.402	60	10.50
30 ℃	0.828	0.833	0.800	0.750	0.333	60	8.27

在Morris模式中,S_i、F、P_F、$P_♀$均是I值的组成成分,除了F是人为规定标准的产卵量外,其余组分都由于不同温度而发生变化,从而也引起I值的变化。从表3看出,20 ℃和30 ℃下的I值分别只占25 ℃下I值的33.52%和78.76%,而20 ℃下的I值又只占30 ℃下I值的42.56%,为了说明各个组分

变化对 I 值的影响，可以应用庞雄飞（1982）提出的组分分析公式予以讨论。

$$M(a_i) = 1 - \frac{a_i}{s_i}$$

$$M(a_1, a_2, \cdots, a_n, a_{P_F}, a_{P_♀}) = \sum_{i=1}^{n}(1 - \frac{a_i}{s_i}) \times (1 - \frac{a_{P_F}}{P_F}) \times (1 - \frac{a_{P_♀}}{P_♀})$$

式中 $M(a_i)$ 表示第 i 个组分增加死亡率后的 I 值为原 I 值的倍数，$M(a_1, a_2, \cdots, a_n, a_{P_F}, a_{P_♀})$ 表示在每个组分上分别增加一个死亡率 $a_1, a_2, \cdots, a_n, a_{P_F}, a_{P_♀}$ 后的 I 值为原 I 值的倍数。

按以上公式比较计算三种温度下的 $M(a_i)$ 值和 $M(a_1, a_2, \cdots, a_n, a_{P_F}, a_{P_♀})$ 的值如表4。

表4　三种温度 I 值的组分分析表

I 值变化	温度比较		
	25 ℃和20 ℃	25 ℃和30 ℃	30 ℃和20 ℃
$M(a_1)$	0.921 9	0.885 6	1.001 1
$M(a_2)$	0.936 5	0.928 7	1.008 4
$M(a_3)$	0.889 3	1.066 7	0.833 8
$M(a_{P_F})$	0.530 0	0.828 4	0.639 6
$M(a_{P_♀})$	0.825 1	1.058 0	0.761 3
$M(a_1, a_2, a_3, a_{P_F}, a_{P_♀})$	0.335 7	0.768 9	0.409 9

从表4不仅可以看出不同组分变化对 I 值的影响，而且还可从中得出温度对平均产卵量的作用是影响桔全爪螨 I 值变化的主要原因的结论。例如以 25 ℃和 20 ℃比较，$M(a_{P_F})=0.53$，这意味着由于 20 ℃下达到标准产卵量的比率（0.213）较 25 ℃下达到标准产卵量的比率（0.402）低 0.189，所以 20 ℃下的 I 值只有 25 ℃下 I 值的 53%。

（2）桔全爪螨实验种群的 Leslie 矩阵

要了解桔全爪螨实验种群随时间变化的数量动态以及不同时间种群结构的年龄组配，仅仅应用前述生命表及 Morris 模式是无能为力的，而 Leslie 矩阵却为此提供了简便的方法。

为了满足 Leslie 矩阵时齐性的要求，首先按桔全爪螨各发育阶段的平均历期（表5），以一天为单位划分时间间隔，如果该螨在第 i 个发育阶段的存活率为 S_i，并假设各时间间隔内的存活率（S_{ij}）相等，则，

$$S_{ij}=S_i^{\frac{1}{m}}(i=1, 2, \cdots, n\text{ 发育阶段})$$

（$j=1, 2, \cdots, m$ 第 i 个发育阶段划分的时间间隔）

其中成螨期各时间间隔的存活率，$S_{n1}, S_{n2}, \cdots, S_{nm}$ 均为1，这是因为成螨死亡的信息，已经反映在雌雄性比和雌螨平均产卵量的资料中。

例如在20 ℃下卵期存活率 $S_1=0.862$，根据卵的平均历期将其划分为8个时间间隔，则每个时间间隔的存活率 $S_{11}=S_{12}=\cdots=S_{18}=0.862^{\frac{1}{8}}=0.9816$，而20 ℃下，成螨期分为6个时间间隔，则每个时间间隔的存活率 $n_{41}=n_{42}=\cdots=n_{46}=1$。

表5 桔全爪螨各发育阶段平均历期

单位：d

发育阶段	温度		
	20 ℃	25 ℃	30 ℃
卵	8	5	4
幼螨	2	1	1
若螨	6	6	5
雌成螨	7	6	4
世代	23	18	14

同理，如果桔全爪螨的繁殖力为 f（$f=FP_FP_♀$），产卵期划分为 k 个时间间隔，假设各时间间隔的繁殖力相等，则各时间间隔的繁殖力为 f/k，例如在20 ℃下雌螨的繁殖力 $f=12.75\times0.571=7.28025$，根据产卵期（除去产卵前期）将其划分为6个时间间隔，则每个时间间隔的繁殖力为 7.28025/6=1.213375。

按照以上计算结果，即可代入矩阵 M，同时令各发育阶段相继时间间隔的种群数量所构成的与 M 的同维列向量为 N，则Leslie矩阵有如下形式：

$$\begin{pmatrix} 0 & \cdots & 0 & 0 & \cdots & 0 & f/k & \cdots & f/k & f/k \\ s_{11} & \cdots & 0 & 0 & \cdots & 0 & 0 & \cdots & 0 & 0 \\ \vdots & & \vdots & \vdots & & \vdots & \vdots & & \vdots & \vdots \\ & & s_{1m} & 0 & \cdots & 0 & 0 & \cdots & 0 & 0 \\ & & & s_{21} & \cdots & 0 & 0 & \cdots & 0 & 0 \\ & & & & & \vdots & \vdots & & \vdots & \vdots \\ & & & & & s_{2m} & 0 & \cdots & 0 & 0 \\ & 0 & & & & & \vdots & & \vdots & \vdots \\ & & & & & & s_{n1} & & 0 & 0 \\ & & & & & & & & \vdots & \vdots \\ & & & & & & & & s_{nm} & 0 \end{pmatrix} \times \begin{pmatrix} n_{11/0} \\ n_{12/0} \\ \vdots \\ n_{21/0} \\ n_{22/0} \\ \vdots \\ n_{31/0} \\ \vdots \\ n_{n2/0} \\ \vdots \\ n_{nm-1/0} \end{pmatrix} = \begin{pmatrix} n_{11/1} \\ n_{12/1} \\ n_{21/1} \\ n_{22/1} \\ n_{31/1} \\ n_{n2/1} \\ n_{nm-1/1} \end{pmatrix}$$

显然以上矩阵可写成如下形式：

$N_1 = MN_0$

$N_2 = MN_1$

$\vdots$

$N_t = MN_{t-1}$

按照上式，$N_0=10$ 粒（初孵卵），在电子计算机上模拟 40 d，作如图 1 的桔全爪螨种群整体曲线和分部曲线。

种群整体曲线代表按照时间序列（天）算出的各发育阶段个体的总数（对数坐标），种群分部曲线分别代表现存卵、幼螨、若螨和成螨的个体数量（仍用对数坐标）。图 1 表明，在三种恒温条件下，随着温度升高，世代周期缩短，在 45 d 中，20 ℃下种群出现 1 次完整的高峰，25 ℃下出现 2 次，而 30 ℃下却出现 3 次，在较高温度下，由于世代增多，各代增长的速率加快，从而使桔全爪螨种群达到较高水平。但是在三种不同恒温条件下，该螨种群逐代增长的趋势却是大体一致的，并且整体曲线的高峰总是与各代卵的高峰相吻合。当某个时间只存在一种发育阶段的螨态时，则种群整体曲线必然和该螨态的分部曲线相重合，相反，整体曲线和分部曲线重合越少，就意味着种群世代和各发育阶段之间重叠越明显。例如 25 ℃下，在第 34 和 35 天都有四个螨态同时存在，而 20 ℃下，一天中最多只出现三个螨态。

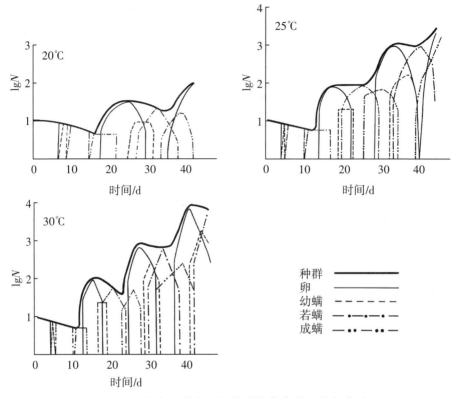

图1 不同温度下桔全爪螨种群整体曲线和分部曲线

(3)桔全爪螨在25 ℃下的生殖力表

前述桔全爪螨实验种群生命表,完全把种群数量变化按离散世代进行处理,这虽然可以提供世代之间种群消长的某些信息,但是却忽略了种群结构及年龄组配对其动态的影响。将生命表数据转换为Leslie矩阵的形式,虽然可以部分地反映种群结构、年龄组配的状况,但是,由于它仅仅采用了各个历期的平均值,而不能反映不同发育阶段及整个世代在历期上的偏差,对于提前发育的个体和时滞个体的信息不能反映。同时,Leslie矩阵对于各历期存活率的处理,是假设同一发育阶段存活率是一致的,而实际上情况当然不会这样简单。桔全爪螨世代明显重叠,无论是自然种群还是实验种群,在某一时刻几乎都可以观察到不同螨态或同一螨态的不同时龄个体。随着时间的推移,部分个体相继死亡,新的个体相继产出,为了阐明与种群结构及年龄组配有关的生殖力与死亡率之间的关系,同时反映时滞个体对种群动态的作用,根据实验室饲养资料,组建25℃下桔全爪螨种群生殖力表,如表6。

表6 桔全爪螨25℃下的生殖力表

x	l_x	m_x	$l_x m_x$	$l_x m_x x$	x	l_x	m_x	$l_x m_x$	$l_x m_x x$
1	1.000	0			22	0.519	1.4	0.726 6	15.985 2
2	0.915	0			23	0.433	1.1	0.476 3	10.954 9
幼期 ⋮	⋮	⋮			24	0.407	1.2	0.488 4	11.721 6
⋮	⋮	⋮			25	0.378	1.1	0.415 8	10.395 0
12	0.830	0			26	0.326	1.1	0.358 6	9.323 6
13	0.815	1.0	0.815 0	10.595 0	27	0.278	1.0	0.270 8	7.311 6
14	0.796	1.5	1.194 0	16.716 0	28	0.248	1.2	0.297 6	8.332 8
15	0.796	1.7	1.353 2	20.298 0	29	0.193	2.2	0.424 6	12.313 4
16	0.748	0.9	0.673 2	10.771 2	30	0.141	1.1	0.155 1	4.653 0
17	0.685	1.5	1.027 5	17.467 5	31	0.093	1.1	0.102 3	3.171 3
18	0.667	1.5	1.000 5	18.009 0	32	0.074	0.5	0.037 0	1.184 0
19	0.637	1.6	1.019 2	19.364 8	33	0.037	1.0	0.037 0	1.221 0
20	0.593	1.7	1.008 1	20.162 0	34	0.037	0		
21	0.544	1.5	0.816 0	17.136 0	Σ			12.704 0	247.218 3

表6中,x表示以日为单位的时间间隔;l_x表示任一个体在x期间得以存活的概率,亦即特定年龄存活率;m_x表示在x期间平均每雌产雌数,亦即特定年龄产雌率。雌性比按实际统计的69.2%计算,假设在x年龄每雌平均产卵数为N_x,则$m_x=0.692N_x$,这样$l_x m_x$即给出了在每一个年龄期间雌螨的生殖数。如果以x为横坐标,以$l_x m_x$为纵坐标,则曲线下的面积$\int_0^\infty l_x m_x \mathrm{d}x$就是该种群在整个存活期间产生后代的总和,这个积分称为净生殖率(R_0)。即:

$$R_0=\int_0^\infty l_x m_x \mathrm{d}x \text{ 或近似地 } R_0=\sum l_x m_x$$

由于螨种群是在最适温25℃的恒温条件下饲养,其食物条件不受限

制,如果再假设种群增长不受自身密度的制约,那么种群将按指数形式增长,即:

$$\frac{dN}{dt}=r_m N, \quad N_t=N_0\exp(r_m t)$$

式中 r_m 表示种群内禀增长能力。

令上式中 $\exp(r_m)=\lambda$,则 $N_t=N_0\lambda^t$, $\lambda=\sqrt{N_t/N_0}$ 或 $\lambda=N_{t+1}/N_t$

由该式看出,λ 恰好表示种群在经过一个单位时间后为原种群数量的倍数,所以称为"周限增长率"。

若种群世代的平均周期为 T,则 $N_T=N_0\exp(r_m T)$ 或 $N_T=N_0\lambda^T$

按照上述 λ 的意义,根据上式,每世代种群增长 $\lambda^T=\exp(r_m T)$ 倍。并且因为有

$R_0=\lambda^T=\exp(r_m T)$,所以 $r_m=\ln R_0/T$

当种群年龄结构稳定,并在一个无限的环境中增长时,有:

$$\int_0^\infty \exp(-r_m x)l_x m_x dx=1$$

假设不同个体的发育速率呈正态分布,则上述方程可逼近为差分的综合,即:

$$\sum_0^\infty \exp(r_m x)l_x m_x=1$$

上式两端同乘 $\exp(6.9078)$,则 $\sum_0^\infty \exp(6.9078-r_m x)l_x m_x=1\,000.047$

应用表6的数据,在电子计算机上代入各种可能的 r_m 值,计算结果,当取 $r_m=0.1403$ 时,$\sum_0^{34}\exp(6.9078-r_m x)l_x m_x=999.2503$

如前所述,还计算得:

$R_0=\sum l_x m_x = 12.704$

$T=\dfrac{\ln R_0}{r_m}=\dfrac{2.5419}{0.1403}=18.1176(d)$

$\lambda=\exp(r_m)=\exp(0.1403)=1.1506$

上述计算结果的生态学意义是,在25 ℃恒温饲养条件下,桔全爪螨的内禀增长能力 r_m(瞬间出生率和瞬间死亡率之差)为0.1403;一个世代的平均时间 T 为18 d,一个世代的增殖倍数为上代的数量的12倍;种群平均每

经过一天为原数量的 1.15 倍。

此外,还可以考虑种群平均增长一倍所需要的时间。要使种群增长一倍,当满足:

$$\frac{N_t}{N_0}=2 \quad 即 \exp(r_m t)=2, t=\ln 2/r_m$$

将 $r_m=0.1403$ 代入,则 $t=4.94(\text{d})$,即每经过 5 d,种群就增长一倍。

这些结果和桔全爪螨实验种群生命表 Leslie 矩阵所反映的信息以及在实验室所观察到的情况是基本吻合的。

讨论与小结

生命表是近年来广泛应用于研究昆虫、螨类种群数量动态的重要方法。本文根据室内饲养资料,组建了 20 ℃、25 ℃和 30 ℃下桔全爪螨实验种群生命表,并通过适当的数据处理,使 Morris 模式和 Leslie 矩阵模型结合起来用于生命表的数据分析,同时组建 25 ℃下的桔全爪螨生殖力表分析了与该螨种群结构及年龄组配有关的生殖力和死亡率之间的关系。

生命表分析表明,温度对桔全爪螨实验种群的数量变化,主要通过对世代总存活率、雌雄性比及产卵量的影响而起作用。其中后两者的综合作用又是影响该实验种群 I 值大小最重要的因子。应用 Leslie 矩阵模型,在电子计算机上进行模拟试验,表明了种群数量及结构随时间序列而变化的不同特点。

实验种群生命表所表明的种群动态和自然种群的真实反应有很大的差别,但它却为研究自然种群的动态以及计算机模拟提供了基础。

原文刊载于《西南农学院学报》,1985 年第 3 期

朱砂叶螨对不同农药抗药性发展趋势的研究

郭凤英,赵志模

摘要:探讨朱砂叶螨室内种群的抗药性发展状况,实验结果表明,氧化乐果、三氯杀螨醇、双甲脒、哒螨灵对于朱砂叶螨种群,只要给予一定的选择压力,经数次用药该螨就会发展对这些杀虫(螨)剂的抗性,且随着用药次数的增加,各品系抗药性呈现上升趋势。在相同条件下,该螨对上述药剂的抗性发展速率不同,4种杀虫(螨)剂抗性发展速率由大到小依次为:氧化乐果、双甲脒、三氯杀螨醇、哒螨灵。

关键词:朱砂叶螨;抗药性;发展趋势

Study on Development Tendency of Pesticides Resistance in *Tetranychus cinnabarinus* (Acari: Tetranychidae)

Guo Fengying, Zhao Zhimo

Abstract: Development of pesticides resistance was studied in *Tetranychus cinnabarinus* (Boisduval). These results showed that the spider mite can develop resistance to omethoate, dicofol, amitraz and pyridaben after spraying several times and resistance factor of every strain was up when the spraying times was increasing. This pesticides-resistance of the mites was different in same conditions. The development of omethoate-resistance had the greatest speed on the spider mite, followed in order by omethoate, amitraz, dicofol, and pyridaben.

Key Words: *Tetranychus cinnabarinus* (Boisduval); pesticide-resistance; development tendency

朱砂叶螨 *Tetranychus cinnabarinus*(Boisduval)为多食性重要农业害螨,分布在世界各温暖地区。我国北至河北、山西,西至甘肃、四川,南至广东、广西、云南,东濒沿海、台湾均有分布。寄主植物有棉花、烟草、玉米、高粱、豆类、瓜类、果树、月季等100多种经济、粮食及观赏植物,棉花受叶螨危害严重时,叶片焦枯脱落,幼龄不能成熟,产量下降,甚至歉收。叶螨自20世纪60年代在我国对内吸磷和对硫磷产生抗药性,随着化学合成杀虫剂在农业生产中的广泛应用,抗药性问题逐渐突出,叶螨对多种杀虫(螨)剂已产生了抗性。笔者选用朱砂叶螨作试螨,对其抗药性的形成和发展趋势进行研究,为叶螨防治中杀虫剂品种的选择提供依据。

害虫抗药性的形成主要是由于杀虫剂选择作用,生物因子和操作因子影响着抗药性的形成和发展,其中操作因子是人为控制的因子,如药剂种类、理化性质、剂型、用药量、用药次数、施药方式和施药的虫期及范围等。本实验采用不断向1个种群连续给予药剂压力的方法,促使叶螨抗药性的快速形成。

1 材料与方法

1.1 试虫饲养

把采自田间的朱砂叶螨移接到盆栽豇豆苗上,在人工气候室内饲养约60代,将这一室内种群分为5个部分,其中一部分不喷药,并被视为敏感品系,另外4部分分别用不同的农药连续处理。叶螨饲养条件为:温度 (26 ± 1) ℃,相对湿度55%～60%,16 h光照(每处理8只40 W日光灯直射),8 h黑暗。

1.2 农药品种

本实验选用的4种杀虫(螨)剂分别是重庆农药厂生产的40%氧化乐果乳油,福建省建瓯农药厂生产的20%三氯杀螨醇乳油,德国先灵公司生产的20%双甲脒乳油和扬州百灵农药厂生产的20%哒螨灵可湿性粉剂。

1.3 抗药性品系的选育方法

从敏感品系开始培育朱砂叶螨的抗药性品系。分别选用上述4种农药,按照下述流程开展抗性选育。以种群死亡率75%～85%的选择压力,长江-08型喷雾器喷洒药剂,喷药24 h后将存活的叶螨个体转移到新的豇豆苗上,在养虫室内饲养,待种群发展到一定数量时再一次喷药。用药2次后,为保证75%~85%的死亡率,要适当提高用药浓度,并用玻片浸渍法做1次毒力测定,计算其LC_{50},了解抗药性的发展。

2 结果与分析

2.1 敏感品系的LC_{50}

用室内饲养2a未与农药接触过的朱砂叶螨对4种杀虫(螨)剂做了毒力测定,其毒力回归直线作为敏感基线,并用此毒力回归线求得该螨对上述4种农药的LC_{50}(表1)。

表1 朱砂叶螨对4种杀虫(螨)剂的敏感基线

杀虫(螨)剂	毒力回归方程	X^2	LC_{50}/(mg·L^{-1})	SE
氧化乐果	$Y=2.2176+1.3434X$	2.406*	117.8047	11.0375
三氯杀螨醇	$Y=1.2721+2.6460X$	4.553*	25.6378	1.2801
双甲脒	$Y=-0.1148+2.2113X$	2.563*	205.6026	15.6107
哒螨灵	$Y=1.5502+1.4088X$	5484*	281.0287	24.3190

注:*显著水平。

2.2 抗药性发展趋势及其模型

朱砂叶螨在上述4种杀虫(螨)剂处理下的毒力测定结果见表2。

表2 朱砂叶螨的抗药性系数

用药次数/次	抗药性系数			
	氧化乐果	三氯杀螨醇	双甲脒	哒螨灵
0	1.0000	1.0000	1.0000	1.0000
2	2.2742	7.4526	2.5422	1.6720
4	3.6402	9.1976	3.5607	2.3486
6	11.9253	15.5877	4.4166	2.5175

续表

用药次数/次	抗药性系数			
	氧化乐果	三氯杀螨醇	双甲脒	哒螨灵
8	24.258 4	16.326 1	4.752 0	3.117 6
10	31.320 5	18.747 1	7.537 7	3.194 2
12	40.009 8	23.371 9	8.226 5	3.537 2
14	45.049 2	21.872 0	18.478 7	3.804 7
16	65.058 8	26.351 1	22.298 0	4.329 1
18	69.143 3	29.251 7	33.546 5	6.081 5
20	88.967 1	30.341 9	35.337 3	7.114 7
22	95.105 3	38.087 7	46.734 0	7.921 6
24	112.504 0	41.313 2	53.153 1	11.132 4
26	120.702 5	42.839 8	60.104 0	—
28	152.832 7	55.585 2	62.610 6	15.669 8

从表2看出,随着用药时间的延长,朱砂叶螨对这4种农药的抗药性系数逐渐升高,表明抗药性由弱到强。朱砂叶螨对氧化乐果有较快的抗性发展速度,经28次用药筛选后抗药性系数达到了152.832 7倍,可能是朱砂叶螨对氧化乐果的抗性是单个不完全显性主基因控制,具有潜在的高发展速度。

在室内饲养的朱砂叶螨经三氯杀螨醇的逐次汰选,抗药性系数呈平缓的增加。第26次用药后的毒力测定,LC_{50}为12.839 8倍,经28次用药,朱砂叶螨对三氯杀螨醇的抗性系数为55.585 2倍,抗性系数1次增加了12.745 4倍,对抗性发展过程的分析和最后1次毒力测定表明的抗性突然增加,说明了该品系对三氯杀螨醇的抗性在未来时期内还会快速升高。

实验结果表明朱砂叶螨对双甲脒也能产生抗性,且随汰选时间的延长而上升。纵观其抗药性发展过程得知,如继续用药,抗药性系数稳中会有缓慢的增加。

朱砂叶螨对哒螨灵抗性的发展相当缓慢,28次用药后,抗药性系数仅达15.669 8倍。

图1中显示了4个品系的抗药性发展趋势。抗氧化乐果品系（Ro）不但抗药性系数最高、发展速度最快，而且从图形可分析出在未来一段时间其抗性还会快速上升，越是到了高抗品系，抗药性发展速度越快。抗双甲脒品系（Ra），在用药开始时抗药性发展速度较慢，第12次用药后抗药性快速增加，24次用药后趋于平缓，未来抗药性发展潜力不大。抗三氯杀螨醇（Rd）的抗性系数低于对氧化乐果的抗性系数，但趋势却很相似，只是每次用药后抗药性增加幅度不如Ro品系的大，未来也有较快速发展的趋势。抗哒螨灵品系（Rp）的抗药性系数增加最慢，未来抗药性也不会有快速的发展。4个品系用药次数和抗药性系数之间的模型如下：

$Y_{Ro} = -15.595\ 3 - 5.211\ 5X \quad r = 0.981\ 1$ ……………（1）

$Y_{Rd} = 2.579\ 4 + 1.626\ 9X \quad r = 0.981\ 5$ ……………（2）

$Y_{Ra} = -9.506\ 5 + 2.113\ 8X \quad r = 0.959\ 1$ ……………（3）

$Y_{Rp} = -0.112\ 0 + 0.131\ 3X \quad r = 0.912\ 1$ ……………（4）

方程（1）—（4）经 F 检验均达极显著水平（$P<0.01$）。从4个回归式的斜率来看，Ro>Ra>Rd>Rp，和抗药性系数一致，抗氧化乐果品系抗性发展速度最快。

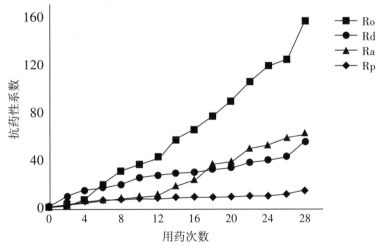

图1 朱砂叶螨对4种农药的抗性发展趋势

3 讨论

本实验对比研究了朱砂叶螨对氧化乐果、三氯杀螨醇、双甲脒、哒螨灵抗药性发展情况。结果是朱砂叶螨对这4种杀虫(螨)剂均能产生抗性,笔者分析可能有2个方面的原因：一是叶螨自身的原因,生殖力强,世代历期短,迁移力弱等,易产生抗药性。二是在选育过程中给予了较大的汰选压力。促使叶螨的抗药性快速形成。在相同实验条件下(包括叶螨的饲养条件),氧化乐果抗性发展最快,其次是双甲脒、三氯杀螨醇,最慢的是哒螨灵。由抗性发展模型得知,各品系的抗药性均随用药次数的增加呈线性增加,但增加速度不同。从抗药性发展状况来看,哒螨灵应该是较理想的杀螨剂,鉴于三氯杀螨醇对天敌昆虫较安全,抗药性发展较慢,也是一种较好的杀螨剂。氧化乐果和双甲脒因朱砂叶螨对其抗性发展较快,笔者认为不宜选作防治叶螨的药剂,和其他杀虫剂轮换使用效果如何有待研究。

原文刊载于《蛛形学报》,1999年第2期

温度对朱砂叶螨二种抗药性品系发育和繁殖的影响

邓新平,何林,赵志模

摘要:朱砂叶螨敏感品系(S)和抗药性三氯杀螨醇品系(Rd)、抗双甲脒品系(Ra)的各螨态历期在一定温度范围内均随温度的增加而缩短,发育速率对温度敏感性的高低依次为 Rd>Ra>S。在相同温度条件下,2个抗药性品系均比敏感品系发育更快。3个朱砂叶螨品系的平均每雌总产卵量与温度的关系,可用二次抛物线方程拟合。在相同温度条件下,2个抗药性品系的生殖力均低于敏感品系,这表明抗药性朱砂叶螨具有生殖不利性。

关键词:朱砂叶螨;抗药性;温度;发育历期;繁殖力

Effect of Temperature on Development and Fecundity of Two Pesticide Resistant Strains of *Tetranychus cinnabarinus* Boisduval (Acari: Tetranychidae)

Deng Xinping, He Lin, Zhao Zhimo

Abstract: The developmental duration and fecundity of dicofol-resistant (Rd) amitraz-resistant (Ra) and susceptible (S) strains of the spider mite, *Tetranychus cinnabarinus* were evaluated at five constant temperatures (16, 21, 26, 31 and 36 ℃). With the range of 16—36 ℃, the developmental period of every life stage (egg, larva, nymph, adult) of the two resistant and susceptible strains were gradually shortened with the temperature increasing. The sensitivity of the developmental rate for temperature was in the order of Rd>Ra>S. At same temperature, the developmental rate of two resistant strain was faster that the susceptible one. The relationship between average number of eggs/female and constant temperature could be fitted to the parabolic equations. Our results suggest that the reproductivity of two resistant strains was lower than the susceptible one at same temperature.

Key Words: *Tetranychus cinnabarinus*; resistant strain; temperature developmental duration; fecundity

朱砂叶螨 Tetranychus cinnabarinus (Boisduval) 为多食性重要农业害螨,分布在世界各温暖地区。寄主植物有棉花、烟草、玉米、高粱、豆类、瓜类、果树、月季等100多种粮食、经济及观赏植物。由于化学农药的大量使用,朱砂叶螨已对多种杀虫(螨)剂产生抗药性。笔者通过室内药剂汰选获得了朱砂叶螨的抗药性品系,研究了2个抗药性品系在不同温度条件下发育历期和繁殖力的变化,以期为叶螨的抗药性治理和新杀虫(螨)剂品种的开发提供理论依据。

1 材料与方法

1.1 供试朱砂叶螨

1.1.1 敏感品系

把采自田间的朱砂叶螨移接到盆栽豇豆苗上,在人工气候室内隔离饲养60代,视为敏感品系(S)。饲养条件为:温度(26±1)℃,55%~60% RH,16 h光照,8 h黑暗。

1.1.2 抗性品系

从敏感品系中分出2部分,分别以三氯杀螨醇和双甲脒进行汰选。在75%~85%的选择压力下,每用药4次,以浸渍法做一次毒力测定,直至获得三氯杀螨醇抗药性品系(Rd)和双甲脒抗药性品系(Ra)。

1.2 毒力测定

采用FAO(1980)推荐的玻片浸渍法,每张玻片的胶带上粘30头健康雌性成螨,放置4 h后镜检,剔除死亡个体,重新补足规定数量。将玻片上粘有雌螨的一端浸入供试药液中,5 s取出,用滤纸吸干附在螨和粘胶表面的药液。每种药液浓度处理3片,同法以清水处理作对照。处理后的玻片放入保湿器内,置于26 ℃,55%~60% RH,16 h光照的培养箱内,24 h镜检死亡数,建立毒力回归方程,计算致死中浓度(LC_{50})。

1.3 发育历期和繁殖力测定

在 16、21、26、31、36 ℃,55%~60% RH,光照 16 h,8 h 黑暗的条件下饲养。1 株豇豆苗叶片上接 1 雌性成螨,产卵后即弃去,每株豆苗保留新鲜卵 1 粒。豆苗外套两端开口的玻璃罩,上覆盖细纱布。每个品系至少饲养 50 头。逐日单头记载叶螨的发育状况,至成螨时,雌雄配对,记录产卵量直到死亡。

2 结果与分析

2.1 朱砂叶螨抗药性品系的选育

表 1 结果表明:朱砂叶螨敏感品系对三氯杀螨醇和双甲脒的 LC_{50} 分别为 25.64 mg/L 和 205.60 mg/L。以三氯杀螨醇汰选的品系,4 次用药后抗药性即开始大幅度增加,汰选 28 次后 LC_{50} 达 1 425.08 mg/L,抗性指数(RF,抗药性系数)为 55.58 倍;以双甲脒处理的品系,抗药性在前期增加较为缓慢,汰选至 12 次后,才开始迅速提高,至 28 次时,LC_{50} 达 12 872.90 mg/L,抗性指数为 62.61 倍。

表 1 朱砂叶螨抗药性品系的汰选结果

药剂	施药次数/次	生物测定时间/d	直线方程(Y=)	LC_{50}/(mg/L)	抗药性系数(RF)/倍
三氯杀螨醇	0	0	1.272 1+2.646 0X	25.64	1.00
	4	80	2.098 6+1.222 9X	235.81	9.20
	8	160	2.125 5+1.096 4X	418.56	16.33
	12	240	1.843 5+1.136 4X	599.28	23.37
	16	320	2.147 4+1.008 1X	675.58	26.35
	20	400	2.206 5+0.966 3X	777.90	30.34
	24	480	1.026 4+1.313 6X	1 059.18	41.31
	28	560	1.311 9+1.169 4X	1 425.08	55.58

续表

药剂	施药次数/次	生物测定时间/d	直线方程($Y=$)	LC_{50}/(mg/L)	抗药性系数(RF)/倍
双甲脒	0	0	$-0.1148+2.2113X$	205.60	1.00
	4	80	$1.5230+1.2138X$	732.08	3.56
	8	160	$0.6570+1.4526X$	976.81	4.75
	12	240	$-0.4425+1.6859X$	1 691.40	8.23
	16	320	$-1.7917+1.8550X$	4 584.52	22.30
	20	400	$-4.1064+2.3584X$	7 265.44	35.34
	24	480	$-4.0601+2.2434X$	10 928.42	53.15
	28	560	$-2.6066+1.8509X$	12 872.90	62.61

2.2 温度对发育历期的影响

在16、21、26、31和36 ℃五种供试温度条件下,朱砂叶螨敏感品系(S)、抗三氯杀螨醇品系(Rd)和抗双甲脒品系(Ra)均能完成其生活史。但各螨态发育历期和世代发育历期均随温度的升高而缩短(表2),并可用一元线性回归方程拟合发育速率V(历期的倒数)与温度T(℃)的关系。其中世代发育速率与温度的回归方程分别为:

S品系:$V=-0.0292+0.00426T$,($r=-0.9958$)

Rd品系:$V=-0.0475+0.00616T$,($r=-0.9957$)

Ra品系:$V=-0.0413+0.00564T$,($r=-0.9956$)

回归方程的斜率在一定程度上反映了发育速率对温度的敏感性。斜率越大,表明温度对发育速率的影响越明显。由上述3个回归方程看出,世代发育速率对温度的敏感性依次为Rd>Ra>S。

在相同温度条件下,除个别温度和螨态组合外(如36 ℃下Ra的幼螨期),抗药性品系的各螨态历期和世代发育历期均较敏感品系的短。在16~36 ℃范围内,世代发育历期S品系为26.76~7.97 d,Rd品系为20.53~5.93 d,Ra品系为20.53~6.12 d。这表明,在相同温度条件下,抗性品系较之敏感品系发育速率加快。表2的结果显示,在16～31 ℃范围内,温度越高,抗性品系加速发育越快。但36 ℃下,这种加速发育的效应有所降低。

表2　不同温度下朱砂叶螨各品系的发育历期

品系	温度/℃	卵期/d	幼螨期/d	若螨期/d	产卵前期/d	产卵期/d	成螨期/d	世代历期/d
S	16	10.13±0.35	5.00±0.00	4.63±0.52	7.00±1.41	25.50±0.98	37.38±7.09	26.76±1.39
S	21	5.55±0.61	4.26±0.75	4.15±0.86	2.29±0.47	25.29±7.31	33.65±7.20	16.25±1.48
S	26	3.88±0.45	1.98±0.36	3.73±0.45	2.15±0.37	22.17±2.63	24.69±2.98	11.74±0.51
S	31	2.91±0.23	1.96±0.20	3.38±0.46	1.91±0.20	12.70±0.12	15.23±0.57	10.16±0.89
S	36	2.89±0.29	1.00±0.02	3.00±0.00	1.08±1.19	8.64±1.63	10.75±1.82	7.97±0.34
Rd	16	8.69±0.85	3.08±0.64	3.38±0.51	5.38±0.96	22.76±7.40	27.77±5.53	20.53±1.33
Rd	21	5.00±0.07	2.52±0.51	3.14±0.36	1.57±0.50	22.04±2.59	25.38±4.46	12.23±2.69
Rd	26	3.00±0.03	2.00±0.01	2.47±0.52	1.29±0.46	19.08±8.20	24.50±2.82	8.76±0.50
Rd	31	2.71±0.47	1.17±0.31	2.00±0.00	0.79±0.26	10.13±3.17	11.33±5.42	6.67±0.26
Rd	36	2.40±0.21	1.00±0.00	2.00±0.00	0.53±0.52	8.14±2.30	8.64±3.71	5.93±0.39
Ra	16	8.17±1.15	3.50±1.30	3.72±0.65	5.14±0.69	22.17±8.96	28.94±8.40	20.53±1.86
Ra	21	5.34±0.55	2.79±0.41	3.61±1.46	1.52±0.69	18.83±6.34	23.73±3.13	13.26±1.11
Ra	26	3.00±0.10	1.60±0.50	3.13±0.35	1.27±0.45	21.30±5.23	22.21±5.43	9.00±0.37
Ra	31	2.61±0.25	1.43±0.30	2.59±0.28	1.19±0.40	10.59±4.64	12.89±2.98	7.82±0.45
Ra	36	2.06±0.25	1.13±0.34	2.00±0.00	0.93±0.18	5.63±2.66	7.88±2.31	6.12±0.62

注:表中数据为$\overline{X} \pm SD$。

2.3 温度对生殖力的影响

在5种供试温度条件下,以26 ℃最适合于朱砂叶螨的产卵繁殖,其平均每雌总产卵量S品系为197.31粒,Rd品系为124.00粒,Ra品系为136.80粒(表3)。平均每雌总产卵量(Y)与温度(T)的关系,可用二次抛物线方程拟合为:

S品系:$Y=-0.967T^2+48.771T-426.857$,($R^2=0.7568$)

Rd品系：$Y=-0.659T^2+33.813T-307.934$，（$R^2=0.9839$）

Ra品系：$Y=-0.790T^2+39.290T-354.702$，（$R^2=0.9839$）

在过低（16 ℃）或过高（36 ℃）温度条件下，平均每雌总产卵量锐减。由于受雌成螨产卵历期的影响，S品系平均每雌日产卵量（表3）由3.58粒增加到10.81粒；Rd和Ra品系在16～31 ℃范围内，随温度增高，平均每雌日产卵量分别由2.59粒增加到9.62粒和由3.38粒增加到9.98粒。但在36 ℃时，两个抗性品系平均每雌日产卵量反而分别下降为7.37粒和5.90粒。

表3 不同温度下抗药性和敏感品系的繁殖力

	温度/℃	S	Rd	Ra
平均每雌总产卵量/粒	16	91.38±3.87	58.85±2.52	74.83±7.33
	21	196.76±7.46	123.00±3.83	113.72±8.22
	26	197.31±4.75	124.00±8.59	136.80±7.45
	31	116.77±9.26	97.43±2.67	105.74±3.47
	36	93.45±8.80	60.00±2.13	33.31±6.75
平均每雌日产卵量/粒	16	3.58±0.15	2.59±0.11	3.38±0.33
	21	7.78±0.30	5.58±0.17	5.34±0.39
	26	8.90±0.21	6.50±0.45	7.27±0.40
	31	9.19±0.73	9.62±0.26	9.98±0.33
	36	10.81±1.02	7.37±0.26	5.90±0.20

注：表中数据为 $\overline{X}\pm SD$。

3 小结

3.1 抗药性品系的选育

经过28次药剂汰选50代后，朱砂叶螨对三氯杀螨醇和双甲脒分别产生了约55倍和62倍的中等抗性。从毒力回归直线的斜率和抗性发展趋势看，2个品系仍处于杂合子阶段，抗性仍有继续发展的潜力。Ra品系前期抗性发展较为缓慢，可能与抗性基因的起始频率太低有关。

3.2 生长发育

朱砂叶螨敏感品系和2个抗性品系的各螨态历期和世代发育历期在16~36℃温度范围内均随温度的增高而缩短。从发育速率与温度的回归方程看出,抗性品系的发育对温度更加敏感。在相同温度条件下,抗性品系较之敏感品系发育更快。这一研究结果与高宗仁等研究朱砂叶螨抗溴氰菊酯、氧化乐果、杀虫脒和三氯杀螨醇4个品系的生物学特性的结论相同。但发育速率加快与不同抗性品系及其不同抗性水平的关系如何,有待进一步研究。

3.3 生殖力

吴益东等、于金凤等、潘文亮等、唐振华等对昆虫的研究结果表明,昆虫产生抗药性后,生殖能力降低,称为抗药性昆虫的生殖不利性。本研究也证实在不同温度条件下,朱砂叶螨抗药性品系的产卵期和产卵量均短于和低于敏感品系,与上述的报道相吻合。但高宗仁等曾报道溴氰菊酯、氧化乐果对朱砂叶螨有刺激增殖作用,笔者认为抗性昆虫、螨类的生殖不利性可能与药剂的种类,昆虫、螨类的抗性机理和抗性水平有关。深入这一领域的研究,将有利于阐明昆虫、农螨的猖獗机制和进行综合防治。

原文刊载于《蛛形学报》,2002年第2期

朱砂叶螨阿维菌素抗性品系选育及适合度研究

何林,杨羽,符建章,王进军,赵志模

摘要:为研究朱砂叶螨抗性适合度,在抗性选育基础上实验室组建了朱砂叶螨相对敏感品系、抗阿维菌素品系在豇豆和棉花上的生命表(26 ℃),比较了相对敏感品系和抗性品系在两种寄主上发育历期、繁殖力及种群参数的差异。结果表明:连续汰选42代,朱砂叶螨对阿维菌素(乳油制剂)产生约8.7倍抗性;与相对敏感品系相比,抗性品系在棉花上若螨期缩短0.55 d、世代历期延长0.88 d,平均产卵量无显著差异;在豇豆上卵期缩短1.71 d,若螨期延长0.41 d,世代历期缩短2.06 d,平均产卵量减少84.29粒/雌。抗性品系在棉花和豇豆上的相对适合度分别为0.89和0.58,表明该螨产生抗药性后,存在适合度缺陷。

关键词:朱砂叶螨;抗药性;相对适合度

Resistance Selection and Relative Fitness of *Tetranychus cinnabarinus* (Boisduval) to Abamectin

He Lin, Yang Yu, Fu Jianzhang, Wang Jinjun, Zhao Zhimo

Abstract: Through resistance selection, the life table of abamectin-resistant and susceptible strains of *Tetranychus cinnabarinus* (Boisduval) on cotton and cowpea were established in laboratory to evaluate the relative fitness of two strains. The results showed that after 42 generations selection. *T. cinnabarinus* developed 8.7-fold resistance to abamectin-EC. Compared with the relative susceptible strain, the duration of nymph of the resistant strain shortened 0.55 d, and the generation time extended 0.88 d when fed on cotton. If fed on cowpea, the duration of egg shortened by l.71 d, nymph stage elongated 0.41 d, the generation time shortened 2.06 d, and the fecundity decreased 84.29 eggs per female. The relative fitness of the abamectin-resistant strains fed on cotton and cowpea were 0.89 and 0.58, respectively. The results also revealed that the abamectin-resistant strain of *T. cinnabarinus* had fitness cost.

Key words: *Tetranychus cinnabarinus* (Boisduval); resistance; relative fitness

适合度(fitness)是以遗传物质为基础的有机体所具有的对生态环境的适应、生存、繁殖的相对能力,并决定了生物有机体在自然选择压力下的行为。当人工选择生物体某一性状时,选择出的个体在生态环境中常常适合度降低。Georghiou 等发现,在进行抗药性选择时,节肢动物的抗性品系与敏感品系相比,前者具有较低的生殖力和较长的发育历期。抗药性害虫在除去药剂选择压的情况下,存在适合度劣势,如发育历期延长、死亡率增加、生殖力减退和越冬能力降低等,在有机氯、有机磷、氨基甲酸酯、拟除虫菊酯或昆虫生长调节剂,甚至生物杀虫剂 Bt 等不同类型药剂产生抗性的多种害虫上均有报道。

朱砂叶螨 *Tetranychus cinnabarinus*(Boisduval)是广泛分布在我国棉花和多种蔬菜上危害严重而又难于防治的一种害螨,因其繁殖力强、世代周期短、活动范围小、近亲交配率高、受药机会多,其抗性问题甚至比其他农作物害虫更为严重。特别是近年来,部分棉区大力推广应用转 Bt 基因棉,虽然棉铃虫的发生危害受到很大抑制,但是一些刺吸式害虫如朱砂叶螨、棉蚜、烟粉虱等危害更显突出。

作者在组建朱砂叶螨敏感品系、抗药性品系分别取食棉花和豇豆后生命生殖力表的基础上,分别研究了朱砂叶螨抗性品系在两种寄主上的相对适合度,以期为该螨的可持续治理提供理论依据。

1 材料与方法

1.1 材料

供试螨类:将采自重庆市北碚区田间豇豆苗上的朱砂叶螨移接到新鲜盆栽豇豆苗上,在人工气候室内饲养 5 年约 100 代左右,并视为相对敏感品系(SS)。从相对敏感品系中分出一个亚品系,采取 1.2 的方法筛选出阿维菌素抗性品系(AbR)。供试寄主:棉花(渝棉 5 号,西南农业大学农学及生命科学学院提供)、豇豆(当地自留种,市场购买)。供试药剂:阿维菌素(abamectin)1.8% 乳油(河南新霸王化工有限公司)。

1.2 抗性选育和生物测定方法

从相对敏感品系开始培育朱砂叶螨的阿维菌素抗药性品系。每个品系的起始代用 F_0 表示,药剂筛选后第 $1,2,\cdots,n$ 代,分别以 $F_1,F_2,\cdots,F_n$ 表示。生物测定方法参照联合国粮农组织(FAO)推荐的玻片浸渍法。

1.3 生命表组建

一株子叶展开的棉花(豇豆)苗在营养钵中栽培。叶上接一供试雌性成螨,该螨产卵后即弃去,每株棉苗保留新鲜卵1粒。棉株外套一两端开口的玻璃罩,上端口覆细纱布,在供试温度(26±1)℃下的光照培养箱内饲养,饲养期间停止用药。敏感和抗性品系至少各饲养50头(即从50粒卵开始)。逐日单头记载叶螨的发育状况,到成螨时,雌雄配对交配,并记录每雌每日产卵数,直至死亡。以赵志模等的方法组建朱砂叶螨敏感品系、抗性品系分别在棉花和豇豆上的生命生殖力表。

1.4 数据处理软件

生物测定中毒力回归方程的计算、生命表各生命参数的计算等分别由IRM、Excel和SPSS10.0等软件完成。

2 结果与分析

2.1 朱砂叶螨对阿维菌素的抗性发展趋势

朱砂叶螨阿维菌素抗药性品系的培育过程见表1。朱砂叶螨经阿维菌素多代筛选后,抗性缓慢而稳定地上升。连续筛选42代,抗性系数为8.65。抗性增加较快的三个阶段为 $F_{10}\sim F_{18}$、$F_{22}\sim F_{26}$ 和 $F_{26}\sim F_{30}$,抗性系数分别增加1.89、2.11和1.36。从抗性增长的趋势及各次毒力测定中毒力回归线的 b 值看,该抗性筛选品系的群体异质性较小(b 值较大),如果继续用药,抗性还会缓慢增长,但出现抗性急增阶段的可能性较小。

表1 朱砂叶螨对阿维菌素的抗性选育

筛选代数	毒力回归线	卡方值(χ^2)	LC_{50}(95%置信限)/(mg·L^{-1})	抗性系数
F_0	$Y=10.794\ 0+3.278\ 9X$	3.01	0.017(0.010~0.024)	1.00
F_4	$Y=10.951\ 6+3.102\ 2X$	2.73	0.012(0.007~0.017)	0.71
F_{10}	$Y=10.842\ 1+3.576\ 1X$	0.66	0.023(0.014~0.032)	1.35
F_{18}	$Y=11.520\ 4+5.182\ 6X$	2.99	0.055(0.053~0.057)	3.24
F_{22}	$Y=9.102\ 5+3.421\ 0X$	0.85	0.063(0.060~0.066)	3.71
F_{26}	$Y=8.472\ 6+3.453\ 8X$	1.25	0.099(0.058~0.140)	5.82
F_{30}	$Y=7.931\ 8+3.213\ 4X$	1.34	0.122(0.116~0.128)	7.18
F_{34}	$Y=3.725\ 4+1.957\ 3X$	0.16	0.132(0.120~0.144)	7.76
F_{42}	$Y=7.544\ 7+3.060\ 4X$	1.34	0.147(0.140~0.154)	8.65

2.2 朱砂叶螨在不同寄主上的发育及生殖情况

朱砂叶螨相对敏感品系(SS)及抗药性品系(AbR)分别在棉花及豇豆上发育和生殖的情况见表2。与SS品系相比,AbR品系在棉花上除若螨期显著缩短、世代历期明显延长外,其他各螨态发育历期、平均产卵量无明显差异;在豇豆上,AbR品系较SS品系的卵期和世代历期均显著缩短,若螨期明显延长,且平均产卵量显著减少。

表2 朱砂叶螨敏感、抗性品系在不同寄主上的发育历期及繁殖力

寄主	品系	卵期/d	幼螨期/d	若螨期/d	产卵前期/d	世代历期/d	平均产卵量/(粒/雌)
棉花	敏感品系	3.23±0.43a	1.09±0.29a	4.55±0.60a	0.55±0.60a	8.86±0.47a	93.45±35.28a
	抗性品系	3.00a	1.00a	4.00b	0.74±0.54a	9.74±0.54b	87.91±26.48a
豇豆	敏感品系	4.36±0.18a	2.16±0.50a	2.94±0.65a	0.45±0.31a	9.94±0.72a	179.82±46.46a
	抗性品系	2.65±0.79b	1.88±0.49a	3.35±0.61b	0.49±0.18a	7.88±1.11b	95.53±44.98b

注:表中数据为平均值±标准差,同一列数据后有相同英文字母表示经Duncan氏多重比较后差异不显著($P\geq0.05$)。

2.3 朱砂叶螨年龄特征生殖力及存活曲线

根据 SS 和 AbR 品系分别在棉花和豇豆上的生殖力和存活率资料组建朱砂叶螨实验种群生命生殖力表。根据生命生殖力表，以朱砂叶螨 SS 品系、AbR 品系分别在棉花和豇豆上的逐日每雌产卵数和特定年龄存活率来表示其在不同条件下的生殖力和存活情况。图 1 表明：朱砂叶螨 SS、AbR 品系取食棉花后，最高日产卵数分别为 10.5 粒和 9.2 粒，抗性品系的生殖力较低（AbR 品系的生殖力曲线在 SS 品系下方），但产卵期延长 5~6 d；在豇豆上，分别具有 12.8 粒和 13.4 粒的最高日产卵数，敏感品系的生殖力明显高于抗性品系（SS 品系的生殖力曲线明显高于 AbR 品系）且产卵期延长 12 d 左右。图 2 表明，朱砂叶螨分别取食棉花和豇豆后，在棉花上抗性品系的存活时间较相对敏感品系延长约 6 d，在豇豆上较相对敏感品系缩短约 12 d；在棉花上抗性朱砂叶螨生命前期存活状况较差，生命后期存活状况较好，而在豇豆上抗性朱砂叶螨的后期存活状况一直较差。

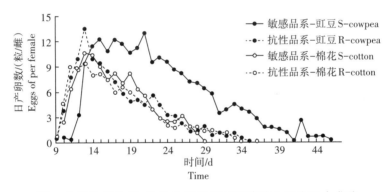

图 1 朱砂叶螨敏感、抗性品系在豇豆和棉花上的生殖力曲线

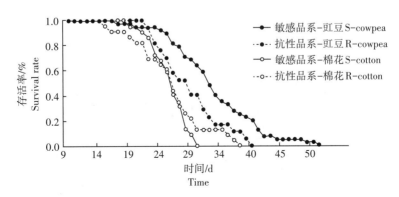

图 2 朱砂叶螨敏感、抗性品系在豇豆和棉花上的存活曲线

2.4 朱砂叶螨实验种群生命参数和相对适合度

根据朱砂叶螨的生命生殖力表,求出两个品系取食不同寄主后的种群生命参数:内禀增长率(r_m=lnRn/T)、净生殖率($Rn=\sum l_x m_x$)、周限增长率($\lambda=e^{r_m}$)、世代平均历期($T=\sum l_x m_x X/Rn$)和种群加倍时间(t=ln2$/r_m$)。从表3可以看出,朱砂叶螨敏感品系、抗性品系分别取食不同寄主后除 Rn 差异较大外,其他种群参数差别不大。以 Rn 的比值表示朱砂叶螨在不同寄主上的适合度(Rf=AbR 的 Rn/SS 的 Rn),相对于 SS 品系,AbR 品系在棉花和豇豆上的适合度分别为0.89和0.58,均小于1,表明抗性品系在两种寄主上均存在适合度缺陷,而在豇豆上适合度缺陷更大。

表3 朱砂叶螨敏感、抗性品系在不同寄主上种群参数及相对适合度

寄主	品系	种群参数					相对适合度
		r_m	T	t	λ	Rn	Rf
棉花	敏感品系	0.281 1	16.327 9	2.465 6	1.324 6	98.517 3	1.00
	抗性品系	0.283 4	15.815 5	2.446 0	1.327 6	88.391 3	0.89
豇豆	敏感品系	0.260 0	19.992 3	2.666 5	1.296 9	180.732 2	1.00
	抗性品系	0.304 2	15.322 2	2.278 8	1.355 5	105.705 9	0.58

3 讨论

一般来说,农药毒力测定和抗性选育要求用有效成分的纯品,以避免商品制剂中多种杂质和农药溶剂对有效成分真正毒力水平的干扰。由于实验条件限制(实验室没有阿维菌素纯品),本研究使用的是1.8%阿维菌素乳油,毒力数据可能不十分准确,但从试验数据看,抗性发展趋势是一致的,即通过连续汰选,朱砂叶螨仍会对生物源农药阿维菌素产生缓慢而稳定的抗性。室内连续筛选42代,朱砂叶螨对阿维菌素(乳油制剂)产生8.65倍抗性,抗性系数不算高,但该抗性筛选品系体内的酶系已发生深刻变化,因此抗性适合度也会发生相应改变,主要表现为卵期、幼螨期发育加快,平均产卵量明显降低。与 SS 品系相比,AbR 品系在两种寄主上(棉花和

豇豆)均存在适合度缺陷。这与前人的抗药性害虫存在生物适合度劣势的研究结果一致。

国外有关螨类的抗药性最早见于1949年美国纽约州棉田使用对硫磷防治棉叶螨失效的事例。国内于1960年初在湖北荆州棉区发现棉红蜘蛛对有机氯农药产生了抗性,并由此开展了对朱砂叶螨的抗性研究。研究抗药性叶螨的适合度,对其抗药性治理(IRM)有着重要的意义。现在对适合度的评价方法尚有不同的看法。Roush等认为在考虑生物潜能(biotic potential,BP)时,平均发育时间(mean development time)的重要性大于生殖力(fecundity,F),因为发育时间(development time,DT)的变化对繁殖潜能的影响要比生殖力的影响大得多,因此提出用BP=$\log_e$F/DT来表示抗性家蝇的适合度。EL-Rhatib等认为净生殖率(Rn)更能代表基因的适合度,因此建议用Rn作为评估各基因适合度的标准。Ferrari在研究抗药性致倦库蚊 *Culex quinquefasciatus* Say 的生殖潜能时认为内禀增长率r_m不能很好地表示品系间的适合度,因为发育时间对r_m的影响要比对Rn的影响大。唐振华等的研究结果表明,淡色库蚊 *Culex pipiens pallens* 抗马拉硫磷(malathion)纯合子(RR)和杂合子(RS)的Rn与敏感纯合子(SS)的差异很明显,Rn和T是生物适合度的两个重要方面,都不能忽视。本研究结果表明,各种群参数中Rn的差异较大,较能代表朱砂叶螨的相对适合度,并且Rn还直接影响着未来时刻叶螨的种群数量,因此适合度的比较采用Rn值。

原文刊载于《植物保护学报》,2004年第4期

黄花蒿提取物对柑橘全爪螨的生物活性

张永强,丁伟,田丽,赵志模

摘要:通过系统测定黄花蒿对柑橘全爪螨的杀螨生物活性,为综合开发利用黄花蒿提供科学依据。采集6月份和7月份的黄花蒿植株,分成根、茎、叶3个部分,采用石油醚(30~60 ℃)、石油醚(60~90 ℃)、乙醇、丙酮和水溶剂的平行和顺序提取方法,获得54种提取物,并测定了它们对柑橘全爪螨的室内杀螨活性。结果表明,7月份的黄花蒿植株的生物活性大多优于6月份,并以7月份叶的丙酮平行提取物活性最高。对柑橘全爪螨处理48 h后,7月份叶的丙酮平行提取物对柑橘全爪螨的LC_{50}(0.422 2 mg/mL)仅为6月份(0.948 9 mg/mL)的44.49%。对黄花蒿7月叶的丙酮提取物进行柱层析得20种不同的组分,其中组分17的杀螨活性最高,与其他组分的杀螨活性存在显著差异($P<0.05$)。

关键词:黄花蒿;柑橘全爪螨;杀螨生物活性

Acaricidal Bioactivity of *Artemisia annua* Extracts Against *Panonychus citri* McGregor (Acari:Tetranychidae)

Zhang Yongqiang, Ding Wei, Tian Li, Zhao Zhimo

Abstract: To determine the acaricidal bioactivities of *Artemisia annua* against *Panonychus citri* McGregor, so to provide a scientific basis for the comprehensive development and utilization of *Artemisia annua*. Collect The whole *A. annua* plants were collected in June and July respectively, and the whole plants were cut into root, stems and leaves, petroleum ether (30—60 ℃), petroleum ether (60—90 ℃), ethanol, acetone and water were used and extracted by the parallel and sequenced solvents extraction methods, finally 54 kinds of extracts were obtained and their indoor acaricidal bioactivity against *Panonychus citri* were determined. The results showed that the biological activity of the *Artemisia annua* plant in July was better than in June and the acetone parallel extract of leaf in July was stronger than other extracts. The LC_{50} of the acetone parallel extract of leaf in July was 0.422 2 mg/mL, which was only 44.49% of 0.948 9 mg/mL in June. With the bio-guided isolation method, the acaricidal activity of different components isolated from *A. annua* July leaf parallel acetone extracts by column chromatography were determined in the laboratory. The 20 kinds of components were obtained, the acaricidal activity of component 17 was the strongest. There was a significant difference between component 17 and other components ($P<0.05$).

Key words: *Artemisia annua*; *Panonychus citri* McGregor; acaricidal bioactivity

0 引言

柑橘全爪螨[*Panonychus citri*(McGregor)]又称柑橘红蜘蛛,是柑橘上普遍发生的最严重的害螨之一,发生代数多,全国各柑橘产区均有分布,为中国柑橘生产的最重要的害虫之一。使用化学农药制剂防治柑橘全爪螨的研究很多,但效果均不理想。黄花蒿(*Artemisia annua*)为菊科蒿属一年生草本植物,因富含青蒿素而闻名于世,对黄花蒿的研究报道主要集中在医药方面。笔者前期的研究表明,黄花蒿对朱砂叶螨[*Tetranychus cinnabarinus*(Boisduval)]具有较强的生物活性。有关植物提取物防治柑橘全爪螨的研究已有相关报道,如韩建勇等研究了白花丹(*Plumbago zeylanica* L.)根提取物对柑橘全爪螨的杀螨活性,发现其具有优良的杀螨、杀卵和产卵抑制活性。周顺玉等报道了18种植物的乙醇提取物对柑橘全爪螨的生物活性,其中测试结果表明,七叶一枝花(*Paris polyphylla*)的毒杀效果最强。以上研究结果表明从植物中寻找杀螨活性物质是可行的,但黄花蒿对柑橘全爪螨的生物活性研究未见报道。为了明确黄花蒿对柑橘全爪螨的生物活性,选用西南大学研究院柑橘园区多年未施药区的柑橘全爪螨为供试对象,试图寻找对柑橘全爪螨活性较高的提取物,并探讨了活性较强的提取物不同柱层析组分的杀螨活性。为提取植物源的生物活性物质,开发新的柑橘害螨控制药剂提供科学依据。

1 材料与方法

1.1 材料

1.1.1 供试螨类

柑橘全爪螨[*Panonychus citri*(McGregor)]:采自西南大学研究院多年未施药的柑橘园区,选择整齐一致的雌成螨作为供试对象。

1.1.2 供试植物材料的来源及提取方法

6月25日和7月25日,在重庆市北碚区西南大学教学实验农场附近,采集黄花蒿全株。将采得的黄花蒿分成根、茎、叶3部分,置于60 ℃烘箱烘干,小型粉碎机粉碎,过80目筛。石油醚Ⅰ、石油醚Ⅱ、乙醇、丙酮和水的平行和顺序膏状提取物的提取方法参见参考文献。

1.1.3 黄花蒿杀螨活性成分的分离

层析用硅胶(100~200目)(青岛海洋化工),干法装柱,称取活性最高的提取物4 g,加4 g硅胶拌匀,加于硅胶柱顶端,用石油醚:丙酮(13:1,11:1,9:1,7:1,5:1,3:1,1:1,1:2,1:3)洗脱,控制洗脱剂流速在300 mL·L^{-1},每40 min收集1份,共分离得48份物质。然后用薄层层析[石油醚:丙酮(3:1)混合液为展开剂]检查,根据 *Rf* 值的大小合并相同部分,共得到20份不同的组分。

1.2 试验方法

取适量的顺序和平行提取物加入一定量丙酮和吐温20,用水稀释配制成5 mg·mL^{-1},作为供试药液。作毒力回归分析时,在初试的基础上选用5~7个浓度。参照FAO推荐的测定螨类抗药性的标准方法——玻片浸渍法并加以改进。结果进行方差分析,并用Duncan新复极差法比较各处理间的效果差异,毒力回归式按机率值分析法计算,由SAS软件统计完成。

2 结果与分析

2.1 黄花蒿6月份和7月份提取物对柑橘全爪螨的触杀活性

用5 mg·mL^{-1}黄花蒿6月份和7月份的不同溶剂的提取物,在实验室条件下测定各自对柑橘全爪螨的触杀活性,结果如图1和图2所示。

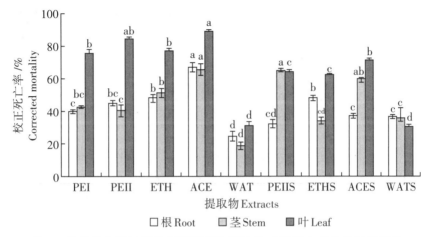

图1 6月份黄花蒿不同溶剂不同部位提取物对柑橘全爪螨的触杀活性
(5 mg·mL^{-1},48 h)

注:PEI为石油醚30~60 ℃平行提取物;PEⅡ为石油醚60~90 ℃平行提取物;ETH为乙醇平行提取物;ACE为丙酮平行提取物;WAT为水平行提取物;PEIIS为石油醚60~90 ℃顺序提取物;ETHS为乙醇顺序提取物;ACES为丙酮顺序提取物;WATS为水顺序提取物。图上所标小写字母表示同一植株部位不同溶剂提取物之间的生物活性差异,相同字母间不存在显著差异,不同字母间存在显著差异($P<0.05$)。下同。

从图1看出,6月份除石油醚60~90 ℃顺序提取物和水顺序提取物外,其他溶剂提取物的叶提取物活性均比根茎提取物的活性高。其中叶的丙酮平行提取物对柑橘全爪螨在5 mg·mL^{-1}处理48 h后,校正死亡率达88.84%。其他的叶提取物对柑橘全爪螨的校正死亡率大多在30%~77%,而水的效果最差。根的活性跟茎相差不多,随提取使用溶剂的不同活性亦有较大变化。

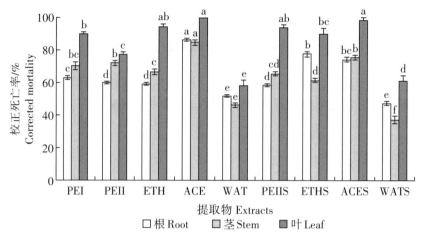

图 2 7月份黄花蒿不同溶剂不同部位提取物对柑橘全爪螨的触杀活性
(5 mg·mL^{-1}, 48 h)

从图2看出,7月份叶的不同溶剂提取物对柑橘全爪螨的生物活性明显高于根和茎的活性,其中叶的丙酮、石油醚Ⅰ、乙醇平行提取物和石油醚Ⅱ、丙酮顺序提取物对柑橘全爪螨的校正死亡率均在90%以上。叶的丙酮平行提取物活性最高,对柑橘全爪螨的校正死亡率达100%。

2.2 黄花蒿几种提取物对柑橘全爪螨的毒力回归分析

分别选择了黄花蒿6月份和7月份对柑橘全爪螨杀螨活性最高的叶丙酮平行提取物为测试材料,测定了它们对柑橘全爪螨的毒力,结果见表1。从表1中看出,黄花蒿7月份叶丙酮平行提取物对柑橘全爪螨的生物活性明显高于6月份,对柑橘全爪螨处理48 h后,7月份的LC$_{50}$(0.422 2 mg·mL^{-1})仅为6月份(0.948 9 mg·mL^{-1})的44.49%。从其LC$_{95}$来看,7月份叶的丙酮平行提取物是3.879 6 mg·mL^{-1},6月份为11.255 6 mg·mL^{-1},也就是说7月份叶丙酮平行提取只需6月份34.47%的用药量就可达到相同的效果。

表1 6月份和7月份黄花蒿叶平行提取物对柑橘全爪螨的毒力(48 h)

月份	毒力回归直线	相关系数	LC$_{50}$及其95%置信区间/(mg·mL^{-1})	LC$_{95}$及其95%置信区间/(mg·mL^{-1})
6	$y=1.859\ 4+1.054\ 9x$	0.968 4	0.948 9(0.601 6~1.496 8)	11.255 6(8.552 6~14.225 8)
7	$y=0.516\ 6+1.707\ 6x$	0.973 2	0.422 2(0.348 3~0.511 7)	3.879 6(3.200 8~4.702 3)

2.3 黄花蒿7月叶丙酮提取物柱层析所得不同组分对柑橘全爪螨的杀螨活性比较

采用生物活性追踪法测定黄花蒿7月叶的丙酮提取物柱层析所得不同组分对柑橘全爪螨的杀螨活性,结果见表2。从中可以看出,在最终分离出的20种组分中,组分17的杀螨活性最高,处理48 h,校正死亡率达到95.65%,与其他组分的杀螨活性存在显著差异($P<0.05$);组分8的活性次之,低于17组分,但均高于其他组分的杀螨活性,与其他组分生物活性间亦存在显著差异。对活性较高的第17和第8组分进行了毒力回归分析,结果列于表3。从中看出,处理48 h,组分17和组分8对柑橘全爪螨的LC_{50}分别为 0.252 6 mg·mL^{-1}和0.393 3 mg·mL^{-1},而7月叶丙酮提取物的LC_{50}为 0.422 2 mg·mL^{-1},组分17的活性相对于叶丙酮提取物而言提高大约1.6倍;而组分8对柑橘全爪螨处理48 h,其LC_{50}为0.393 3 mg·mL^{-1},与提取物本身的LC_{50}相差不大。组分17对柑橘全爪螨生物活性较高,有待进一步深入研究。

表2 黄花蒿7月叶丙酮平行提取物柱层析组分对柑橘全爪螨杀螨活性 (5 mg·mL^{-1}, 48 h)

不同组分	校正死亡率/%		校正死亡率95%置信区间/%	
	48 h	72 h	48 h	72 h
1	31.53 ghi	52.34 efg	27.34~35.71	51.05~53.64
2	35.68 g	55.19 e	32.65~38.70	54.35~56.02
3	58.10 d	66.90 d	54.00~62.19	61.47~72.32
4	59.96 d	79.31 c	53.00~66.92	75.49~83.14
5	77.98 c	85.31 b	77.12~78.84	81.03~89.59
6	26.96 ij	46.70 h	22.69~31.22	42.97~50.44
7	29.61 hij	49.35 fgh	24.66~34.55	44.28~54.42
8	84.02 b	93.61 a	74.75~93.29	88.31~98.92
9	77.35 c	80.66 c	70.52~84.19	77.61~83.71
10	25.73 j	48.47 gh	20.35~31.11	44.04~52.89

续表

不同组分	校正死亡率/%		校正死亡率95%置信区间/%	
	48 h	72 h	48 h	72 h
11	36.57 g	57.27 e	32.79~40.36	52.32~62.23
12	34.32 gh	55.30 e	21.08~47.56	41.28~69.33
13	49.70 f	70.12 d	41.63~57.77	61.07~79.16
14	35.46 g	54.61 ef	23.25~47.67	43.61~65.61
15	31.99 ghi	52.17 efg	26.96~37.01	42.82~61.52
16	26.34 j	46.64 h	22.77~29.91	43.01~50.26
17	95.65 a	98.85 a	95.42~95.89	95.42~102.68
18	75.38 c	86.63 b	62.79~87.98	80.74~92.52
19	55.00 de	76.43 c	49.53~60.46	71.09~81.76
20	50.75 ef	71.60 d	40.86~60.65	60.52~82.69

注：同列数据后标有不同小写字母表示差异显著（$P<0.05$，Duncan新复极差测验）。

表3　黄花蒿7月叶丙酮平行提取物和分离的第17和第8组分对柑橘全爪螨的毒力回归分析

处理	毒力回归直线	相关系数	LC_{50}及其95%置信限/($mg \cdot mL^{-1}$)	LC_{95}及其95%置信限/($mg \cdot mL^{-1}$)
提取物	$y=0.516\ 6+1.707\ 6x$	0.973 2	0.422 2(0.348 3~0.511 7)	3.879 6(3.200 8~4.702 3)
第17组分	$y=0.341\ 2+1.939\ 2x$	0.977 4	0.252 6(0.218 1~0.292 5)	1.780 8(1.537 9~2.062 1)
第8组分	$y=1.538\ 7+1.334\ 0x$	0.925 1	0.393 3(0.313 7~0.493 1)	6.726 5(5.365 2~8.433 2)

3 讨论

鉴于笔者前期工作的研究，已经证实黄花蒿6月份和7月份植株表现出对朱砂叶螨的杀螨活性相对较高，为避免重复工作，本文仅报道了这两个月份黄花蒿植株对柑橘全爪螨的生物活性。7月份黄花蒿植株的不同部位表现出较6月份强的生物活性，单就植株的不同部位来说，叶的提取物活性最强,研究结果与张永强等报道的对朱砂叶螨的生物活性相似。7月叶丙酮平行提取物对柑橘全爪螨在 5 mg·mL^{-1} 处理 48 h 后,校正死亡率达 100%。

同一植物的不同生长期,植物经受的环境条件不同,其次生代谢产物的种类和含量可能不同,7月份是黄花蒿营养生长的关键时期,其自身合成的以保护自身生长而不被外界有害生物为害的次生代谢产物含量也会随之升高。并且朱砂叶螨和柑橘全爪螨均是叶螨科的重要植食性害螨,其为害方式相似,寄主范围重叠严重。所以7月份黄花蒿叶的丙酮平行提取物表现出对朱砂叶螨和柑橘全爪螨较强的生物活性。而有关其毒理学证据有待进一步的试验考证。随后,对活性最高的黄花蒿7月份叶丙酮平行提取物进行柱层析分离,结合详尽的生物测定,在最终分离出的20种组分中,组分17的杀螨活性最高,处理48 h,校正死亡率达到95.65%,与其他组分的杀螨活性存在显著差异($P<0.05$)。组分17对柑橘全爪螨的 LC_{50} 为 0.252 6 mg·mL^{-1},而7月叶丙酮提取物的 LC_{50} 为 0.422 2 mg·mL^{-1},组分17的活性相对于叶丙酮提取物而言提高大约1.6倍。组分17对柑橘全爪螨生物活性较高,有待进一步深入研究。这一结果为进一步从黄花蒿中筛选、分离和鉴定杀螨活性成分奠定了基础,为开发利用中国丰富的黄花蒿资源,寻找柑橘全爪螨防治新药剂具有重要的参考价值。

4 结论

研究表明,7月份黄花蒿植株的不同部位表现出较6月份强的生物活性,单就植株的不同部位来说,叶的提取物活性最强。7月叶丙酮平行提取物对柑橘全爪螨在 5 mg·mL^{-1} 处理 48 h 后,校正死亡率达100%。对活性最高的黄花蒿7月份叶丙酮平行提取物采用柱层析分离出20个组分,其中组分17的杀螨活性最高,处理48 h 和 72 h 后,校正死亡率分别达到95.65%和98.85%,与其他组分的杀螨活性存在显著差异($P<0.05$),且组分17对柑橘全爪螨的 LC_{50} 仅为 0.252 6 mg·mL^{-1},因此组分17对柑橘全爪螨生物活性较高,其具体的杀螨活性物质有待进行深入研究。

原文刊载于《中国农业科学》,2009年第6期

甲氰菊酯和阿维菌素对柑橘全爪螨的亚致死效应

何恒果,闫香慧,王进军,赵志模

摘要:通过叶碟饲养的方法,利用生命表技术,研究了甲氰菊酯和阿维菌素亚致死剂量 LC_{20} 处理柑橘全爪螨若螨后,对试验种群当代(F_0)和后代(F_1、F_2 代)生长发育及繁殖的影响。结果表明:甲氰菊酯 LC_{20} 处理若螨后,当代雌成螨产卵量显著增加;F_1、F_2 代的产卵前期缩短,后代雌性比例增大,且均与对照差异显著;同时,F_1 和 F_2 代种群内禀增长率(r_m)和周限增长率(λ)增大,世代历期(T)和种群加倍时间(D_t)缩短,且 F_2 代与对照相比差异显著。用阿维菌素 LC_{20} 处理若螨后,当代种群雌成螨产卵量显著下降;F_1 和 F_2 代的产卵量也显著下降,但后代雌性比例增大,产卵前期显著缩短;F_1 和 F_2 代的种群 r_m 和 λ 增大,T 和 D_t 缩短,且 F_2 代比 F_1 代更为明显。总体来看,甲氰菊酯和阿维菌素亚致死浓度 LC_{20} 对柑橘全爪螨的影响并不完全相同,甲氰菊酯能够促进当代种群的发展,而阿维菌素对当代种群有一定的抑制作用;但两种杀螨剂亚致死浓度处理柑橘全爪螨对后代种群都有一定的促进作用。研究结果对柑橘全爪螨综合防治策略的制定有一定的指导意义。

关键词:柑橘全爪螨;甲氰菊酯;阿维菌素;亚致死效应;生命表参数

Sublethal Effects of Fenpropathrin and Avermectin on *Panonychus citri*(Acari: Tereanychidae)

He Hengguo, Yan Xianghui, Wang Jinjun, Zhao Zhimo

Abstract: A leaf disc bioassay was employed to examine the effects of fenpropathrin and avermectin with a sublethal concentration of LC_{20} on the development and reproduction of F_0, F_1 and F_2 generations by means of life tables. The results showed that after the treatment of fenpropathrin at the sublethal concentration, the number of eggs laid per female significantly increased in F_0 generation, the pre-oviposition duration was significantly shortened and the female ratio of offspring significantly increased both in F_1 and F_2 generations. The intrinsic rate of increase (r_m) and finite rate of increase (λ) values all increased, and the generation time (T) and population doubling time (D_t) were shortened in F_1 and F_2 generations, with significant difference observed between F_2 generations and the control. After exposure to avermectin, the number of eggs laid per female significantly decreased in F_0 generation, and progeny (F_1 and F_2) also produced fewer eggs than the control, while the female ratio of offspring increased both in F_1 and F_2 generations and the preoviposition period was significantly shortened. The r_m and λ values all increased, and the T and D_t were shortened in F_1 and F_2 generations. Such effects were more obvious on the F_2 generation than the F_1 generation. Generally, the effects of fenpropathrin and avermectin with a sublethal concentration of LC_{20} were not exactly the same on *P. citri*. Fenpropathrin could promote the development of the contemporary population, while avermectin had certain inhibition on the contemporary population, but both played a certain role in facilitating the development of future populations, which was of significance in developing integrated pest management strategies.

Key words: *Panonychus citri* McGregor; fenpropathrin; avermectin; sublethal effect; life-table parameters.

杀虫(螨)剂施用于田间后,除了对昆虫(螨)的直接杀死作用以外,随着个体间接触药量的差异以及时间的推移,对部分个体还存在着亚致死效应,包括害虫生物学和生态学行为的改变、生殖力的变化、抗药性的发展等。亚致死效应的研究能全面了解药剂使用后害虫的存活、发育、生殖及种群增长率的变化,进而有助于害虫管理措施的制订。

近年来,有关亚致死剂量对昆虫(螨)生长发育和繁殖的研究较多,这是探讨抗药性产生的生态学机制之一。Nandihalli等研究发现,亚致死剂量的溴氰菊酯、氯氰菊酯和氰戊菊酯能引起棉蚜(*Aphis gossypii*)的再猖獗。Wang研究发现,LC_{25}吡虫啉和氰戊菊酯可以刺激蚜虫的生殖增长。陶士强等研究发现,朱砂叶螨[*Tetranychus cinnabarinus*(Boisduval)]经亚致死剂量LD_{35}的毒死蜱处理后单雌产卵量显著降低。Marcic等研究发现亚致死剂量的螺螨甲酯和螺螨乙酯都会降低二斑叶螨(*Tetranychus urticae*)生殖,但Landeros等研究发现,LC_{10}亚致死剂量的阿维菌素可促进二斑叶螨的增殖。柑橘全爪螨[*Panonychus citri*(McGregor)]是为害柑橘最为严重的世界性害螨,当前化学防治仍是控制其为害的重要手段之一。甲氰菊酯(fenpropathrin)作为一种广谱高效的拟除虫菊酯类杀虫、杀螨剂,由于其具有高效、低毒、在土壤中不易移动、难以挥发等特点,且可以较好地防治柑橘叶蛾、凤蝶等多种害虫,尤其在害虫、害螨并发时,可虫螨兼治,因此,在柑橘园内仍然广泛使用。阿维菌素(avermectin)为一种高效生物杀虫、杀螨、杀线虫剂,具有高选择性和高安全性,被农业部推荐为高毒农药的首选替代品种之一,在柑橘园内广泛使用。

有研究表明,柑橘全爪螨对甲氰菊酯和阿维菌素抗药性发展很快,对其抗药性产生的生化机理也有涉及,但有关甲氰菊酯和阿维菌素使用后柑橘全爪螨的存活、发育、生殖及种群增长率的变化还未见报道。在LC_{10}—LC_{25}中,常常选择LC_{20}来评估害虫或害螨的亚致死剂量,如张曼丽等用LC_{20}亚致死剂量考察阿维菌素和噻螨酮对刺足根螨(Rhizoglyphus echinopus)休眠体形成与解除的影响。为此,本试验也选取了甲氰菊酯和阿维菌素亚致死代表剂量LC_{20}来处理柑橘全爪螨若螨,考察用药对试验种群当代(F_0)和后代(F_1、F_2代)生长发育及繁殖的影响,以探讨其抗药性产生的生态学原因,为柑橘全爪螨的抗性治理、抗性预防提供理论依据,为柑橘全爪螨的综合治理提供理论指导。

1 材料与方法

1.1 试验材料

供试柑橘全爪螨采自中国农业科学院柑桔研究所的种质资源圃枳壳苗上,在温度为(26±1)℃、相对湿度(RH)为70%~75%、光照14 h和黑暗10 h的人工气候室内用盆栽枳壳苗饲养,饲养至15代后使用,饲养期间未接触任何药剂。

供试药剂92%的甲氰菊酯原药和93%的阿维菌素原药由四川省药品检验所提供。

1.2 试验方法

1.2.1 亚致死浓度的确定

生物测定采用带螨叶片浸渍法。先将原药用丙酮溶解稀释20倍,在预试验的基础上再用蒸馏水稀释5~7个浓度。将新鲜叶片用打孔器打成5 cm直径的叶碟,放入垫有吸水海绵和棉花的培养皿内,并加浅水形成弧岛状,以防叶螨外逃和叶片干枯。随机从室内枳壳苗上分别挑取大小一致、行动活泼的若螨40头于叶碟上,待其稳定后,将带螨叶片用镊子夹住浸入药液中,5 s后取出,迅速用吸水纸吸掉螨体及叶片上多余的药液,放回上述的培养皿内,置于温度为(26±1)℃、相对湿度(RH)为70%~75%、光照14 h和黑暗10 h的人工气候室中,24 h后检查结果,计算死亡率。另以丙酮水溶液(丙酮=0.20%)为对照,对照组死亡率在10%以内为有效试验,所得数据在Excel和SPSS上进行处理,求出毒力回归直线方程和致死中浓度LC_{50}值以及亚致死浓度LC_{20}值。

1.2.2 亚致死剂量及对当代(F_0代)的影响试验

随机从室内枳壳苗上挑取大小一致、行动活泼的桔全爪螨雌成螨200头左右于叶盘上,让其产卵4 h后挑出,产下的卵发育至若螨时用作试验对象。根据形态特征和蜕皮次数,试验中留取若螨150头左右,将带螨叶片浸入预先配好的LC_{20}(1.240 mg·L^{-1})的甲氰菊酯药液或LC_{20}(0.001 mg·L^{-1})的阿维菌素药液5 s,取出,迅速吸掉螨体及叶片上多余的药液,放入之前的培养皿内。24 h后,检查死亡情况并将螨移至新鲜叶片上,继续观察其发育情况及产卵情况(考察对当代F_0的影响)。每处理设置3个重复,以不用药处理作为对照。

1.2.3 亚致死剂量对子代种群生长发育的影响试验

处理后存活的F_0雌成螨达到产卵高峰期时留取卵约100粒(F_1代)观察其孵化情况。待卵孵化后,每天记录蜕皮及存活情况,至成螨后,记录雌雄比例并每天记录产卵数直至成螨死亡。需注意的是,在成为若螨后,观察间隔时间改为8 h,便于确定产卵前期。每处理设置3个重复,以清水处理作为对照。在F_1代产卵高峰期,将雌成螨转移到新鲜叶片让其产卵4 h后,留取所产的卵约100粒组建F_2代的生命表。其余方法同F_1代。

生命表构建参照丁岩钦的方法,采用SPSS 12.0软件对统计数据进行方差分析。

2 结果与分析

2.1 亚致死浓度的确定

根据生物测定结果,将校正死亡率转换为几率值并与药剂浓度的对数值进行回归分析,建立毒力回归方程(表1)。由表1可知,甲氰菊酯和阿维菌素对柑橘全爪螨均有较强的毒力,其中阿维菌素毒力显著强于甲氰菊酯(甲氰菊酯 LC_{50} 为 5.197 $mg·L^{-1}$,阿维菌素 LC_{50} 为 0.020 $mg·L^{-1}$)。甲氰菊酯的 LC_{20} 为 1.240 $mg·L^{-1}$,阿维菌素的 LC_{20} 为 0.001 $mg·L^{-1}$,确定为亚致死剂量。

表1　甲氰菊酯和阿维菌素对柑橘全爪若螨的毒力

药剂	回归方程	LC_{50}(95%置信限)/($mg·L^{-1}$)	相关系数(r)	卡方值(χ^2)	LC_{20}/($mg·L^{-1}$)
甲氰菊酯	$Y=1.352x+4.032$	5.197(3.907~6.706)	0.9806	0.372	1.240
阿维菌素	$Y=0.658x+6.113$	0.020(0.012~0.030)	0.9875	0.459	0.001

2.2 甲氰菊酯和阿维菌素亚致死浓度 LC_{20} 处理对柑橘全爪螨 F_0 代的影响

甲氰菊酯和阿维菌素亚致死浓度处理对柑橘全爪螨试验种群影响较大,若螨受药后四处逃逸,雄螨由于迅速逃逸被水淹死而导致其寿命缩短,但雌成螨逃逸现象不明显,其总体寿命也未受到明显影响。

从表2可以看出,甲氰菊酯和阿维菌素亚致死浓度处理柑橘全爪螨若螨后,成螨羽化率分别为89.7%和91.2%,略低于对照,但三者间无显著差异。用药后对当代雌成螨产卵量影响很大,对照每雌产卵量为42.1粒,甲氰菊酯处理后产卵量显著增多,每雌产卵量达到65.6粒,阿维菌素处理后当代产卵量显著下降,仅为每雌20.1粒。甲氰菊酯处理后所产后代的雌性比例有所下降,阿维菌素则对后代雌性比影响不大。两种药剂使当代成螨平均寿命有所下降,但差异未达到显著水平。总体来看,甲氰菊酯 LC_{20} 对柑橘全爪螨当代种群有刺激增殖的作用,阿维菌素 LC_{20} 对柑橘全爪螨当代种群有一定的负面影响。

表2 甲氰菊酯和阿维菌素亚致死剂量对当代成螨的影响

处理	羽化率/%	每雌产卵量/粒	雌性比/%	产卵前期/d	成螨寿命/d
CK	94.4±0.2a	42.1±8.6b	69.8±0.1a	1.74±0.09a	8.24±0.09a
FF_0	89.7±0.7a	65.6±3.0a	60.1±2.7b	1.34±0.06a	8.37±0.22a
AF_0	91.2±0.7a	20.1±2.8c	63.2±3.8ab	1.56±0.26a	7.68±0.01a

注：CK表示对照；FF_0、AF_0分别表示甲氰菊酯和阿维菌素处理的当代若螨；同列不同小写字母表示差异显著（$P<0.05$）。下同。

2.3 甲氰菊酯和阿维菌素亚致死剂量对柑橘全爪螨子代种群的影响

2.3.1 甲氰菊酯和阿维菌素亚致死剂量对柑橘全爪螨未成熟期发育历期的影响

从表3可以看出，用甲氰菊酯和阿维菌素LC_{20}处理若螨后，柑橘全爪螨F_1和F_2代的成熟前各历期与对照有所不同，两种处理方式之间趋势相同。两种药剂处理后的子代种群（F_1和F_2代）卵期与对照相比有所延长，但差异不显著。甲氰菊酯处理后，F_1代的幼螨期缩短，F_2代无显著变化；若螨期延长，但与对照相比无显著差异。阿维菌素处理后的子代种群（F_1和F_2代）幼螨期无显著变化；但F_1代若螨期显著缩短，F_2代却稍有延长。甲氰菊酯和阿维菌素处理后，其子代种群（F_1和F_2代）的产卵前期均显著缩短，F_1和F_2代的产卵前期分别为0.97和1.04、0.91和0.84 d，均显著低于对照的1.74 d。

表3 甲氰菊酯和阿维菌素亚致死剂量对柑橘全爪螨未成熟期发育历期的影响

处理	卵期	幼螨期	若螨期	产卵前期
CK	5.75±0.58a	1.36±0.28a	2.47±0.16a	1.74±0.09a
FF_1	5.90±0.12a	1.09±0.06b	3.11±0.00a	0.97±0.06b
FF_2	5.88±0.03a	1.37±0.05a	2.89±0.02a	1.04±0.17b
AF_1	5.99±0.02a	1.40±0.01a	1.99±0.04b	0.91±0.06b
AF_2	5.86±0.02a	1.30±0.05a	2.70±0.00a	0.84±0.17b

注：FF_1和AF_1、FF_2和AF_2分别表示甲氰菊酯和阿维菌素处理若螨期的第一代和第二代。下同。

2.3.2 甲氰菊酯和阿维菌素亚致死剂量对柑橘全爪螨产卵量、成螨寿命及存活率的影响

从表4可以看出,甲氰菊酯和阿维菌素处理,对后代的产卵量、成螨寿命及存活率等的影响显著。甲氰菊酯亚致死浓度处理后,F_1代卵孵化率显著降低,F_2代无明显变化;成螨羽化率在子代种群(F_1和F_2代)中无显著变化。每雌产卵量和成螨寿命在两代(F_1和F_2代)之间差异较大,F_1代的每雌产卵量有所增加,成螨寿命也有所增加,F_2代却均有所减少,但都与对照间无显著性差异。处理后子代(F_1和F_2代)雌性比例均显著提高。

阿维菌素亚致死浓度处理后柑橘全爪螨F_1和F_2代的卵孵化率、成螨羽化率都未受到明显影响,但产卵受到抑制。与对照相比,每雌产卵量均显著下降。此外,阿维菌素LC_{20}处理后的子代(F_1和F_2代)雌性比例增大,F_2代比F_1代更为明显,成螨寿命均缩短。

表4 甲氰菊酯和阿维菌素亚致死剂量对柑橘全爪螨卵孵化率、成螨羽化率、每雌产卵量、雌性比和成螨寿命的影响

处理	卵孵化率/%	成螨羽化率/%	每雌产卵量/粒	雌性比/%	成螨寿命/d
CK	97.2±0.3a	94.4±0.2a	42.1±8.6a	69.8±0.1c	8.24±0.09a
FF_1	88.2±0.5b	91.8±1.1a	52.0±1.7a	74.9±0.7b	8.71±0.01a
FF_2	98.1±2.4a	95.2±1.0a	39.8±1.4a	81.9±2.1a	7.27±0.15a
AF_1	98.1±0.1a	97.5±0.3a	21.5±3.2b	70.9±7.5c	7.71±0.01a
AF_2	95.6±3.6a	95.4±2.8a	26.7±3.0b	82.9±2.3a	7.20±0.05a

柑橘全爪螨种群从第10日开始产卵,至第12、13日达到产卵高峰期,至第19日,有的种群每雌产卵量降至0(AF_1)。其中,用药后的后代种群(FF_1、FF_2;AF_1、AF_2)开始出现死亡的时间都有不同程度的提前,但FF_1存活率(l_x)在第33天为0,比对照延迟了4 d,其余各种群与对照相比种群存活率为0的时间均提早(图1)。用甲氰菊酯处理后,虽然F_1代每雌产卵量、后代雌性比例和成螨寿命增加,但是由于其卵的孵化率显著降低,导致其每雌日均产卵量(m_x)并未显著增加(表4、图1);而F_2代虽然成螨每雌产卵量有所降低、寿命缩短,但由于卵孵化率未受影响,成螨羽化率有所增加,产卵前期显著缩短,后代雌性比例显著增大,使其种群的每雌日均产卵量(m_x)

显著增加。阿维菌素用药后,虽然 F_1 和 F_2 代卵孵化率和成螨羽化率无显著差异,每雌总产卵量下降,但由于产卵前期显著缩短,后代雌性比例显著增加,每雌日均产卵量(m_x)在 F_2 代却增加(图1)。

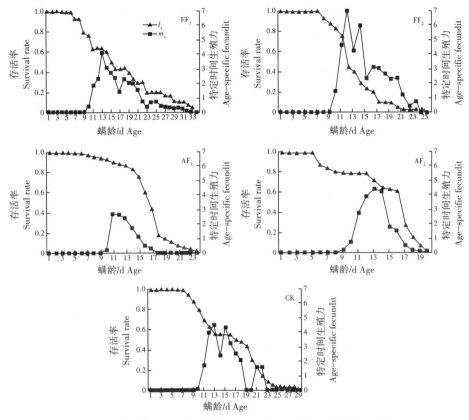

图1 甲氰菊酯和阿维菌素亚致死剂量处理下柑橘全爪螨
的存活率及特定时间生殖力

2.3.3 甲氰菊酯和阿维菌素亚致死剂量对柑橘全爪螨种群参数的影响

从表5可以看出,甲氰菊酯和阿维菌素对柑橘全爪螨试验种群生命表参数影响较大,且两者之间存在一定的差异。甲氰菊酯 LC_{20} 处理柑橘全爪螨后,其 F_1 和 F_2 代种群内禀增长率和周限增长率增大,世代历期和种群加倍时间缩短;F_2 代与对照相比差异显著。与对照相比,阿维菌素 LC_{20} 处理后,F_1代的净增殖率减小,F_2代的净增殖率无明显变化;F_1和F_2代的内禀增长率增大、周限增长率增大,世代历期和种群加倍时间缩短,且F_2代比F_1代更为显著。

表5 甲氰菊酯和阿维菌素LC_{20}对柑橘全爪螨生命表参数的影响

处理	净增殖率(R_0)	内禀增长率(r_m)	周限增长率(λ)	世代历期(T)	种群加倍时间(D_t)
CK	14.03±1.01a	0.167 1±0.007 8b	1.181 9±0.009 3c	15.80±0.31a	4.16±0.20a
FF_1	14.21±1.15a	0.183 3±0.005 6ab	1.203 8±0.003 4b	14.42±0.32a	3.76±0.31b
FF_2	13.98±0.09a	0.230 2±0.014 9a	1.245 3±0.008 7a	11.91±0.04b	3.29±0.17c
AF_1	9.73±1.04b	0.179 3±0.000 3ab	1.196 4±0.009 0bc	12.68±0.21b	3.87±0.13ab
AF_2	14.11±1.00a	0.205 1±0.006 4a	1.227 6±0.000 1ab	12.91±0.04b	3.38±0.34c

3 讨论

本研究中,用甲氰菊酯LC_{20}处理F_0代若螨后,不仅当代每雌产卵量显著增多,且F_1、F_2代的雌性比增大,净增殖率、内禀增长率增大,世代历期缩短,种群加倍时间缩短。所以总体来看,甲氰菊酯LC_{20}对柑橘全爪螨试验种群有一定的刺激增殖作用,这种有利于种群发展或刺激种群增长的现象被称为低剂量刺激效应(hormoligosis)。以往的研究表明,使用拟除虫菊酯类农药后,柑橘全爪螨普遍具有再猖獗现象,原因之一就是拟除虫菊酯类农药对该螨具有刺激增殖的作用。其他杀螨剂对柑橘全爪螨也有相类似的结果,研究发现,使用马拉硫磷后,柑橘全爪螨的每雌产卵量、净增殖率都显著增加。此外,以前的报道也表明甲氰菊酯亚致死浓度对二斑叶螨和山楂叶螨(Tetranychus viennensis)具有类似的刺激增殖作用。这些结果与本试验结果一致。阿维菌素属生物类杀螨剂,对成螨、若螨和幼螨均高效。本试验用阿维菌素LC_{20}处理若螨后,对柑橘全爪螨试验种群也有极大的影响,且影响具有复杂性。它可以显著降低当代和后代(F_1、F_2)雌成螨的产卵数,使试验种群成螨寿命缩短,在短期内有抑制种群数量的作用,这与Flores等对二斑叶螨、李定旭等对山楂叶螨的研究结果相似。但子代(F_1、F_2)种群内禀增长率增大,种群加倍时间缩短,F_2代与对照相比差异显著。有研究指出,生命表参数是评价药剂对昆虫种群全面影响的重要方法,其中内禀增长率 r_m 是推测种群增长或降低的重要参数。从这点来看,阿维菌素 LC_{20} 对柑橘全爪螨未来种群具有潜在的生殖促进作用。Landeros 等的研究结果显示,在 LC_{10} 剂量下,阿维菌素对二斑叶螨种群有一定的刺激作用,表现为种群内禀增长率增大,这个结果与本试验结果较为一致。但谷清义等研究发现,LC_{20}、LC_{10} 的阿维菌素作用于土耳其斯坦叶螨(*Tetranychus turkestani*)成螨后,其次代种群生殖力降低,这与本试验结果不同。

对农药的评价,不能仅仅比较作用虫态短期内的死亡率,还至少应该考虑该虫态能够成功变态进入下一虫态的存活率,并且还要综合评价整个种群所受到的影响。研究表明,有的农药对害虫具有极强的后效性,不仅对当代害虫有较强的致死作用,还可影响到第 2 代、第 3 代的生长与繁殖。本研究结果中亚致死浓度的甲氰菊酯和阿维菌素对柑橘全爪螨种群影响较大,对该螨的影响不只在当代,还影响其第 2 代,对第 3 代的影响甚至更为明显,这也证实了农药对害虫(螨)的亚致死效应不仅仅停留在当代,对子代的作用可能更影响整个种群的发展动态。因此,农药对害虫(螨)子代种群的影响也是评价农药对害虫亚致死效应时应该考虑的必要因素。

研究表明,受药时期不同则农药对害虫的影响存在差异,如 Marcic 用四螨嗪处理二斑叶螨早期卵(0~24 h)和晚期卵(72~96 h)后,对其生长发育的影响存在显著性差异,前者促进生殖,后者抑制生殖。另外,甲氰菊酯亚致死剂量可以减少害螨捕食性天敌塔六点蓟马(*Scolothrips takahashii*)的产卵量和寿命,阿维菌素亚致死剂量对捕食螨黄瓜钝绥螨(*Amblyseius cucumeris*)、伪新小绥螨(*Neoseiulus fallacis*)均具有负面影响,这从某种程度上也促进了害螨种群的增长。由此可见,农药的亚致死效应与杀虫(螨)剂种类、剂量、用药时期、害虫(螨)种群以及杀虫(螨)剂对害虫(螨)天敌的影响等因素都有关。

原文刊载于《应用生态学报》,2016 年第 8 期

Effects and Mechanisms of Simulated Acid Rain on Plant-Mite Interactions in Agricultural Systems. I. The Direct Effects of Simulated Acid Rain on Carmine Spider Mite, *Tetranychus cinnabarinus*

Zhao Zhimo, Luo Huayuan, Wang Jinjun

Abstract: The effects of simulated acid rain sprayed directly on carmine spider mite, *Tetranychus cinnabarinus* (Boisduval), were investigated in a series of laboratory trials. The indirect effects, by the induced changes of development, physiological and biochemical aspects of host plant were eliminated from this experiment. The results indicated that there was no acute lethality of sprayed simulated acid rain of 3.0≤pH≤4.0 on the tested developmental stages of carmine spider mite. However, the developmental duration of larvae, nymph and pre-oviposition was prolonged; and the longevity of adult mite and the duration of generation were shortened. The results of the life table study showed that the mite, which was sprayed directly by the simulated acid rain of pH=4.0, pH=3.5 and pH=3.0, had lower values of the net reproduction rate (R_0), innate rate of increase (r_m), finite rate of increase (λ) and mean generation duration (T), and greater population doubling time (t) than controls. The results implied that the population development and reproduction of the mite were restrained by direct application of acid rain (3.0≤pH≤4.0)

Key words: *Tetranychus cinnabarinus* (Boisduval); acid rain; direct effect

Introduction

Acid rain is a natural precipitation containing some acidic substances, and its pH values are often below 5.6. The threat of acid rain to the global biosphere has became an environmental problem of worldwide importance. Increased acidity of many lakes and soils, alteration of the bio-communities' equilibria, and declining of forests have been shown to be related to acid rain.

The effects of acid rain on vegetation, including the direct (primary) and indirect (secondary) effects through the influence on soils, have been well documented. However, studies on direct effects of acid rain on phytophagous insects, and indirect effects through the influence on the host plant of insects are relatively limited. Stinner et al. reported that after the corn plants were treated by the simulated precipitation (pH=2.8 and pH=4.2), the food utilization of black cutworm larva [*Agrotis ipsilon* (Hufnagel)] was increased, and the development time was shortened. Braun and Fluckiger found that the population development of *Phyllaphis fagi* (L.) and *Aphis fabae* Scopoli was reduced obviously by acid rain ($2.6 \leqslant pH \leqslant 3.6$). Averill observed that the fecundity of *Rhagoletis pomonella* (Walsingham) decreased on the host plants which were sprayed by the simulated acid rain. In Europe, the infestation of *Scolytus* sp. in forests was intensified because of the acid rain pollution. In China, Li and Wu observed that under the impact of acid rain ($3.0 \leqslant pH \leqslant 4.2$), the honey production of *Apis cerana* Fabricus was reduced by 60% and the mortality increased, while the bees were all killed when the pH values were below 3.0. Yang and Ji, Yang et al. and Ji et al. reported that the number of insect species and population densities of secondary insect pests of masson pine increased significantly in areas subjected to acid rain. They thought this was mainly due to the increased amino acid content of the trees and reduced resistance to insect attack. Chen et al. found restrained development of larva of *Pieris rapae* L. sprayed directly by simulated acid rain (pH=4.0), although the larva fed on the Chinese cabbage which was sprayed by the same acid rain had enhanced development.

Chongqing is a city of the most serious acid rain in China. Based on the Bulletin of Environmental Protection (EP) which was released by Chongqing EP Bureau, the frequency of acid rain increased from 25% to 100% and acidity (pH values) declined from 4.65 to 3.94 during 1981—1987. The serious impact of acid rain in Chongqing has caused soil acidification, forest declining, crop yield reduction, and intensified insect pest infestation. Up to now, little is known about the effects of acid rain on mites. To evaluate the effects and mechanisms of impact of acid rain on plant-insect (mite) systems, the direct effect and indirect effect of acid rain on mite development and reproduction have been investigated since 1994. The following studies were conducted as the first step in examining the direct effect of acid rain on the carmine spider mite, *Tetranychus cinnabarinus* (Boiduval).

Materials and methods

Compound of simulated acid rain

Using the mean values of major components of acid rain recorded from Tongyuan Monitoring Station, Chongqing, the simulated acid rain was prepared by the method of Wang and Qin. The stock solution of H_2SO_4 was added to bring the initial solution to pH=5.6, pH=4.0, pH=3.5 and pH=3.0. Two control treatments were also set up using non-ionic water (pH=6.8, CK1) and natural rain in Chongqing (pH=5.6, CK2).

Mite culture

The strain of carmine spider mite tested in this study was obtained from a laboratory colony established from adults collected on a cowpea bean plant in 1992 in Beibei, Chongqing. This colony was cultured on kidney bean plants for more than two years before use.

The acute lethal effects of acid rain on carmine spider mite

Petri dishes (Diameter 10 cm), each with a water-soaked foam plastic (diameter 8 cm) inside, were used as observation units. The water-soaked foam plastic was covered with a layer of filter paper on which a kidney leaf was placed. The leaf edges were encircled with absorbent cotton to prevent mites from escaping. All leaves used in this study were the same size and from the same phyllotaxis. More than 30 eggs, deutonymphs and female adults were introduced into the dishes, respectively. The simulated acid rain (pH=4.0, pH=3.5 and pH=3.0) and CK (CK1 and CK2) solution with three replicates each treatment were applied to the dishes from a portable overhead sprinkler system. The spray volume for each dish was 2 mL. The dishes were kept at 27 ℃, 80%~90% RH, with photoperiodic regime of L∶D=15 h∶9 h (800—1 000 lx) for three days, and the mortality counts were made daily.

The direct effects of acid rain on carmine spider mite analyzed by the life table method

The basic design of this experiment was similar to the above one. Each kidney bean leaf was divided into 6 sections using absorbent cotton. Two mated female adults were introduced into each section, and allowed to lay eggs. After 12 h, the adults were removed and only one egg was kept in each section. The simulated acid rain was sprayed once at the stages of egg, larvae and nymph, respectively. After spraying, the mites were transferred to unpolluted fresh leaves immediately to eliminate the indirect effects on the mite by the host plant. Each treatment consisted of 20 dishes contained 6 eggs each dish. The development and survival of the mites were observed every 12 hours until all mites died. The life and fertility table were constructed.

Results

The acute lethal effects of acid rain on carmine spider mite

The results (Table 1) indicated that the corrected mortalities due to the simulated acid rain (pH=4.0, pH=3.5 and pH=3.0) and natural rain (pH=5.6) sprayed directly to eggs, nymphs, larvae and female adults were all below 2%, and no significant differences were detected among these treatments ($P \geqslant 0.05$). This suggested that the acid rain of pH$\geqslant$3.0 had no acute lethality to the mites (Table 1).

Table 1. The acute lethal effects of the simulated acid rain to carmine spider mite, *Tetranychus cinnabarinus* (Boisduval)

Treatments	No. of tested mites	Dead mites after 72 h	Mortality/%	Corrected mortality/%
pH=6.8 (CK1)	Egg(196)	2	1.02	
	Nymph(93)	1	1.08	
	Adult(95)	1	1.05	
pH=5.6 (CK2)	Egg(179)	4	2.23	1.22
	Nymph(91)	2	2.20	1.13
	Adult(106)	3	2.73	1.70
pH=4.0	Egg(156)	4	2.56	1.56
	Nymph(94)	2	2.13	1.06
	Adult(106)	3	2.83	1.80
pH=3.5	Egg(118)	3	2.54	1.54
	Nymph(93)	2	2.15	1.08
	Adult(116)	3	2.59	1.56
pH=3.0	Egg(168)	4	2.38	1.37
	Nymph(95)	2	2.11	1.04
	Adult109)	3	2.75	1.72

Note: There was no significant difference of the corrected mortalities among treatments ($P \geqslant 0.05$).

The direct effects of acid rain on the development of carmine spider mite

The results of this experiment are summarized in Table 2. Statistical analysis (Duncan's test) showed that the mites sprayed directly by the non-ionic water (pH=6.8, CK1) and natural rain (pH=5.6, CK2) had similar developmental parameters, including developmental duration, longevity of female adults, period of generation, ratio of sterile for female mite, oviposition period and fecundity ($P \geqslant 0.05$). However, compared with CK1 and CK2, the mites, which were sprayed by the simulated acid rain (pH=4.0, pH=3.5 and pH=3.0), had prolonged duration of larva, nymph and pre-oviposition period as well as the increased percentage of sterile female mites with increasing acidity. But the longevity of female mites, the period of generation, oviposition period and the numbers of egg laid per female were reduced markedly. This implied that the simulated acid rain of pH≤4.0 sprayed directly had the restraining effects on development and especially on the reproductive capability of the mite.

Table2. The direct effects of the simulated acid rain on the development of the carmine red mites (M±SE)

Treatments	CK1, pH=6.8	CK2, pH=5.6	pH=4.0	pH=3.5	pH=3.0
Egg/d	4.41±0.50	4.41±0.49	4.40±0.49	4.40±0.34	4.41±0.35
Larva/d	2.26±0.44	2.26±0.43	2.47±0.50	2.53±0.50	2.62±0.49
Nymph/d	4.02±0.52	4.02±0.40	4.08±0.27	4.23±0.60	4.39±0.62
Pre-ovi/d	1.13±0.34	1.13±0.26	1.24±0.36	1.26±0.37	1.48±0.36
Female longevity/d	9.27±1.19	9.22±1.92	7.42±1.62	6.94±1.52	6.43±1.59
Generation period/d	19.95	19.90	18.37	18.10	17.84
Sterile ratio/%	2.44	2.48	3.13	3.33	6.12
Oviposition period/d	8.33±1.48	8.24±1.47	6.58±1.31	4.75±1.10	4.53±1.11
Egg No. laid per female	44.73±18.10	43.44±16.31	34.25±10.92	24.00±10.92	22.25±9.51

Note: duration=d.

The effects of acid rain on the survivorship and fecundity of carmine spider mite

The age-specific survival rate (l_x) and the female offspring laid per female (m_x) of the carmine spider mite under the direct impact of the simulated acid rain are shown in Table 3.

Although the values of l_x of different treatments had similar trend, the values of l_x at specific ages differed significantly. For instance, on the 6th day, l_x of

CK1 and CK2 were up to 93%—95%; l_x of three acid rain treatments was around 91%. On the 10th day, l_x of CK1 and CK2 were around 70%; while l_x of pH=4.0, pH=3.5 and pH=3.0 was only 57%, 45% and 40%, respectively. On the 15th day, l_x of CK1 and CK2 were about 30%; but l_x of acid rain treatments were all below 20%; the pH=3.5 and pH=3.0 treatments were only 11% and 13%, respectively. These results showed that the direct treatment of simulated acid rain of pH≤4.0 had an unfavorable effect on the survivorship of the carmine spider mite.

Table 3. The population life and fertility table of carmine spider mite, *Tetranychus cinnabarinus* (Boisduval), on the impact of the simulated acid rain

Age(x,d)	pH=6.8		pH=5.6		pH=4.0		pH=3.5		pH=3.0	
	l_x	m_x	l_x	m_x	l_x	m_x	l_x	m_x	l_x	m_x
0	1.000 0		1.000 0		1.000 0		1.000 0		1.000 0	
⋮	⋮		⋮		⋮		⋮		⋮	
6	0.945 9		0.925 9		0.913 6		0.914 0		0.914 5	
7	0.837 8		0.827 2		0.790 1		0.741 9		0.812 0	
8	0.810 8		0.753 1		0.666 7		0.677 4		0.589 7	
9	0.743 2		0.691 3		0.604 9		0.623 7		0.547 0	
10	0.700 3		0.691 3		0.570 4		0.451 6		0.406 8	
11	0.682 1	1.742 1	0.674 1	1.758 3	0.535 8	2.718 9	0.451 6	1.666 7	0.394 9	0.800 9
12	0.675 4	3.898 9	0.674 1	3.937 8	0.518 5	4.809 6	0.421 5	3.392 9	0.394 9	2.770 6
13	0.567 6	6.132 1	0.570 7	5.714 4	0.449 4	5.000 1	0.346 2	3.074 6	0.335 0	4.311 3
14	0.404 0	3.876 3	0.380 2	3.928 7	0.432 1	3.200 1	0.195 6	3.131 9	0.299 1	2.457 2
15	0.321 6	2.214 3	0.311 0	1.666 7	0.207 4	1.798 5	0.105 4	0.918 4	0.131 6	1.428 6
16	0.202 4	6.734 8	0.242 0	6.836 9	0.121 0	6.224 6	0.075 3	3.875 2	0.095 7	4.732 2
17	0.202 4	4.796 0	0.207 4	4.702 5	0.069 1	3.571 5	0.075 3	2.142 9	0.083 8	0.510 2
18	0.166 2	3.005 8	0.155 6	3.889 0	0.051 9	0.476 2	0.075 3	1.852 4	0.071 8	0.714 3

续表

Age(x,d)	pH=6.8		pH=5.6		pH=4.0		pH=3.5		pH=3.0	
	l_x	m_x	l_x	m_x	l_x	m_x	l_x	m_x	l_x	m_x
19	0.135 0	1.920 1	0.138 3	1.875 0	0.017 3	0.000 0	0.000 0	0.000 0	0.070 0	0.000 0
20	0.037 8	3.476 3	0.034 6	5.357 3	0.000 0				0.024 0	
21	0.010 5	0.357 2	0.000 0	0.000 0					0.012 0	
22	0.000 0	0.000 0							0.000 0	

The m_x values in Table 3 showed that the mites, which were directly sprayed by CK1 and CK2, had triple oviposition peaks. These oviposition peaks appeared successively at the 13th, 16th and 20th day, and the mean numbers of female offspring laid per female were 5.7—6.1, 6.7—6.8 and 3.4—5.3, respectively. However, the mites, that were directly sprayed by three acid rain solution, only had two oviposition peaks at 12—13th and 16th day, and the m_x were only 3.0—5.0 and 3.8—6.2, respectively. These results suggested that the direct treatment of the simulated acid rain not only resulted in suppression of the third oviposition peak, but also resulted in markedly reduced numbers of female offspring.

The effects of acid rain on the life table parameters of carmine spider mite

Population net reproductive rate (R_0), intrinsic rate of increase (r_m), finite rate of increase (λ), population mean generation time (T) and population doubling time (t) are very important population life parameters which reflect the role of increase of a population. Table 4 shows that these parameters did not differ between CK1 and CK2, and the mites treated with the simulated acid rain had lower values of R_0, r_m and λ with increasing acidity compared to CK1 and CK2. Accordingly, t values of the treatments were greater than those of CK1 and CK2. These data imply that the treatment of the simulated acid rain of pH≤4.0 had negative effects on the development, survival and reproduction of carmine spider mite.

Table 4. The effect of simulated acid rain on life table parameters of carmine spider mite

	pH=6.8	pH=5.6	pH=4.0	pH=3.5	pH=3.0
R_0	12.808 1	12.825 3	8.975 5	4.548 0	4.324 6
r_m	0.184 1	0.182 8	0.168 7	0.116 3	0.110 3
λ	1.202 1	1.200 6	1.183 8	1.123 3	1.116 6
t/d	3.768 4	3.791 6	4.108 5	5.959 6	6.283 8
T/d	13.855 4	13.955 1	13.009 6	13.018 8	13.270 7

Discussion

This study measured only the direct effects of acid rain on carmine spider mite, which is the effects on mite development, survivorship and reproduction after the mite body is in contact with acid rain directly. The indirect effects, which occur through the induced changes of developmental, physiological and biochemical aspects of host plant was eliminated in this study. The results showed that the acid rain had no acute lethality to carmine spider mite. This is consistent with the results of Zhang. He found that there is no obviously acute lethality of air pollutants to agricultural insects (mites). But it is not in full agreement with the study of Li and Wu in which the mortality of *Apis cerana* significantly increased under the impact of the acid rain of $3.0 \leqslant pH \leqslant 4.2$. Our results also showed that the treatment of simulated acid rain of $pH \leqslant 4.0$ had negative effects on the development, survivorship and reproduction of carmine spider mite. It is fully consistent with the results of Braun and Fluckiger and Chen et al. on *Phyllaphis fagi*, *Aphis fabae* and *Pieris rapae*. It is expected that future studies on the effects of acid rain on plant-mite interaction in agricultural systems will be very important.

Original article was published in *Systematic & Applied Acarology*, volume 4, 1999

Calculation of Developmental Duration of Mites Reared in Groups Compared to Those Reared in Isolation

Zhao Zhimo, McMurtry James Allen

Abstract: Laboratory studies with the phytoseiid mites, *Euseius hibisci* (Chant) and *E. stipulatus* (Athias-Henriot) were conducted to calculate and analyze the mean duration of each developmental stage when reared in groups. The results show that the calculation method suggested in this paper is suitable for group rearing. The results calculated by this method are basically identical with that of calculation under conditions of rearing isolated individuals, provided that mortality during development is low.

Introduction

Mites, such as the Phytoseiidae and Tetranychidae, develop through egg, larva, protonymph, deutonymph and adult stages. Determining the duration of each development stage has been an important aspect in biological studies of mites. In order to determine the duration of development for phytoseiid mites, most investigators, for example, Charlet and McMurtry, Bounfour and McMurtry, Ferragut et al., and Moraes et al., have adopted the method of rearing isolated individuals from egg to adult and calculating average duration and standard deviation. This method has obvious advantages. Each individual can be followed from egg to adult, regardless of developmental stage overlap of the individuals within a cohort. The sex of each individual can be determined when the mites become adults. The developmental stage, in questionable instances, can be determined by observation of the presence or absence of exuviae. However, this method has certain disadvantages. First, it requires more space in controlled temperature cabinets than the method of rearing in groups. Secondly, in nature the mites usually occur in groups or colonies in the microhabitat. Therefore, a method of rearing mites in groups of several individuals would resemble more closely the natural conditions. For example, the premating behavior of males waiting beside female deutonymphs would not be precluded by isolation of mites until maturity. The objective of the present study was to devise a method to calculate the average duration of each development stage of the cohort.

Materials and methods

Two species of phytoseiid mite, *Euseius hibisci* (Chant) and *E. stipulatus* (Athias-Henriot), used in this study were taken from an insectary culture at the University of California, Riverside. Eggs of these two species were obtained by placing several small clumps of cotton on tiles of rearing units of the stock cultures. Twenty-four hours later, eggs which had been oviposited there were collected and isolated individually in test arenas. The test arenas consisted of excised orange leaves placed upper surface down on foam mats in 20 cm×20 cm stainless steel pans with water. A 4 cm×4 cm leaf area was delineated by 1.5 cm-wide strips of Cellucotton®. A single egg was placed in each arena, with a total of 18 eggs for each of the two species of mite. The experiment was conducted in small modified refrigerators in which temperature was maintained at (25±0.5) °C and photoperiod was 12 h of light and 12 h of darkness. Humidity was not controlled.

To determine the duration of each developmental stage, observation with a dissecting microscope at magnifications up to 120× were made every 12 h during development from egg to adult. Only immature stages were differentiated: egg, larva and nymph. When larvae were first observed, pollen of ice plant, *Malephora crocea* (Jacq.) was scattered on the arena surface as the food for the mites. Whenever a leaf began to deteriorate, it was replaced with a fresh one.

主题三 螨类毒理学研究

Table1. Duration in hours of each development stage of *Euseius hibisci* at 25°C

Observation number	Time of mid observations	middle time	1	2	3	4	5	6	7	8	9	10	11	12	13	14	15	16	17	18	E	L	N	A ♂ ♀
0	0	12	E	E	E	E	E	E	E	E	E	E	E	E	E	E	E	E	E	E	18			
1	12	18	E	E	E	E	E	E	E	E	E	E	E	E	E	E	E	E	E	E	18			
2	24	30	E	E	E	E	E	E	E	E	E	E	E	E	E	E	E	E	E	E	18			
3	36	42	E	L	E	E	E	E	L	E	L	E	L	E	E	E	E	E	E	L	13	5		
4	48	54	E	L	E	E	E	L	L	E	L	E	L	E	E	L	L	E	L	L	9	9		
5	60	66	L	N	L	E	L	N	N	L	N	L	N	L	L	N	L	L	N	N	1	9	8	
6	72	78	L	N	L	L	L	N	N	N	N	N	N	N	N	N	N	N	N	N		4	14	
7	84	90	N	N	N	N	N	N	N	N	N	N	N	N	N	N	N	N	N	N			18	
8	96	102	N	N	N	N	N	N	N	N	N	N	N	N	N	N	N	N	N	N			18	
9	108	114	N	A	N	N	N	A	N	N	N	A	N	N	N	N	N	N	N	A			14	4 4
10	120	126	N	A	N	N	N	A	A	N	A	N	A	A	A	N	N	A	A	A			8	10 7 3
11	132	138	A	A	A	A	A	A	A	A	A	A	A	A	A	A	A	A	A	A				18 9 9
		Sex	♀	♂	♂	♀	♀	♂	♀	♀	♂	♀	♂	♀	♂	♂	♀	♂	♀	♂				
Duration of each stage per individual		E	66	42	66	78	66	54	42	66	42	66	42	66	66	54	54	66	54	42				
		L	24	24	24	12	24	12	24	12	24	12	24	12	12	12	24	12	12	24				
		A	48	48	48	48	48	48	60	60	60	60	48	48	48	60	60	60	60	48				
Total			138	114	138	138	138	114	126	138	126	138	114	126	126	126	138	138	126	114				

Calculating method and results

Eggs which were introduced to the experimental arena s were laid during a 24 h period. Because it is not exactly known when these eggs were oviposited, the mid-point (12 h) was chosen as the time of oviposition. According to the method of Perring et al., and ignoring the short time spent in molting, it was assumed that a new stage of mite appeared at the midpoint of two observation times.

Tables 1 and 2 are the observation records of development for the two phytoseiid mites. Under the condition of rearing isolated individuals, mean durations were calculated according to the formula:

$$\overline{D}=\sum_{i=1}^{n}D_i/n \quad\cdots\cdots(1)$$

where $\overline{D}$ is mean duration of eggs, larvae, nymphs or immatures, D_i is duration of certain stage of ith individual, an n is number of individuals in the test.

Standard deviation is estimated from the equation:

$$\overline{SD}=\sqrt{\sum(\overline{D}-D_i)^2/(n-1)} \quad\cdots\cdots(2)$$

where $\overline{SD}$ is the standard deviation of mean duration. The other terms are the same as equation (1).

To consider how to calculate the mean duration of each stage and its standard deviation under conditions of group rearing, we can regard these 18 individual mites as a colony, as presented in the right side of Tables 1 and 2.

Table 2. Duration in hours of each development stage of *Euseius stipulatus* at 25 ℃

No. of observations	Time of observations	middle time	\multicolumn{18}{c|}{Individual}																			Total				
			1	2	3	4	5	6	7	8	9	10	11	12	13	14	15	16	17	18	E	L	N	A	♂	♀
0	0	12	E	E	E	E	E	E	E	E	E	E	E	E	E	E	E	E	E	E	18					
1	12	18	E	E	E	E	E	E	E	E	E	E	E	E	E	E	E	E	E	E	18					
2	24	30	E	E	E	E	E	E	E	E	E	E	E.	E	E	E	E	E	E	E	18					
3	36	42	—	E	E	E	E	E	E	E	E	E	E	L	E	E	E	E	E	E	16	1				
4	48	54	—	L	L	L	L	L	L	L	L	L	L	N	L	L	L	L				16	1			
5	60	66	—	L	L	L	L	L	L	L	L	N	N	N	L	L	L	N	L			13	4			
6	72	78	—	N	N	N	N	N	N	N	N	—	N	N	N	N	N	N	N				16			
7	84	90	—	N	N	N	N	N	N	N	N	—	N	N	N	N	N	N	N				16			
8	96	102	—	N	—	N	N	N	N	N	N	—	A	N	N	N	N	N	N				14	1	1	

续表

No. of obser-va-tions	Time of obser-va-tions	middle time	Individual																		Total					
			1	2	3	4	5	6	7	8	9	10	11	12	13	14	15	16	17	18	E	L	N	A	♂	♀
9	108	114	—	N	—	N	N	—	N	N	N	N	—	A	A	N	N	—	A	N		10	3	1	2	
10	120	126	—	N	—	N	N	—	N	A	A	A	—	A	A	A	A	—	A	A		4	9	3	6	
11	132	138	—	A	—	A	A	—	A	A	A	A	—	A	A	A	A	—	A	A			13	5	8	
		Sex	♂		♀ ♂		♀ ♀	♂ ♀		♂ ♀ ♂ ♀				♀ ♀													

Duration of each stage per individual	E	54 54 54 54 54 54 54 54 54 54 42 54 54 54 54 54
	L	24 24 24 24 24 24 24 24 24 12 12 12 24 24 24 12 24
	N	60 60 60 60 48 48 48 36 60 48 48 48 48
Total		138 138 138 138 126 126 126 102 126 126 114 126

Using *E. hibisci* as an example, we first calculate the mean duration of eggs as follows: starting with 18 eggs, there were 13 eggs and 5 larvae observed at the 3rd observation (36 h); that is, 5 larvae hatched at a mean of 42 h after they were oviposited. Obviously the total development time of these 5 eggs is 42×5=210 (h). At the 4th observation (48 h), 9 eggs and 9 larvae were observed. It shows that there were 9−5=4 larvae which hatched in the period between this time and the previous observation. The total development time of these 4 eggs is 54×4=216 (h). At the 5th observation (60 h), there was one egg, 9 larvae and 8 nymphs observed, Because eggs must be through the stage of larva to nymph, these new larvae, which appeared in the period between the 4th and the 5th observations, were 9+8−9=8. The total development time of these 8 eggs was 66×8=528 (h). At the 6th observation (72 h), all eggs hatched and the new hatching larva is 4+14−9−8=1. The incubation time of this egg was 78 h. Finally, the sum of all egg development periods divided by the number of hatched eggs gives the mean duration of egg stage ($\overline{D}_E$):

$$\overline{D}_E = \frac{42 \times 5 + 54 \times (9-5) + 66 \times (9+8-9) + 78 \times (4+14-9-8)}{5+4+8+1}$$
$$= 57.33 \text{ h}$$

As stated above, the minimum, maximum and average incubation times of these 18 eggs were 42, 78 and 57.33 h, respectively.

Actually, this method of calculating mean duration of development stage in group rearing is equivalent to that of applying frequency mean to calculate the average duration under conditions of rearing isolated individuals.

The mean duration and standard deviation of egg stage is calculated by following two formulas:

$$\overline{D}_E = \sum f D_E / \sum f \quad \cdots\cdots(3)$$

$$\overline{SD}_E = \frac{\sqrt{\sum f D_E^2 - (\sum f D_E)^2 / \sum f}}{Rf - 1} \quad \cdots\cdots(4)$$

where f is frequency of D_E appearing. Note that f is the number of new increasing larvae in the period between this time and the last observation.

For duration of larva and nymph, the method of calculating is basically the same as the method of calculating the mean duration of eggs. But the mean duration of larvae calculated by this method includes that of eggs, and mean duration of nymphs includes that of eggs and larvae. For this reason, the mean duration of larvae and nymphs calculated by this method must subtract that of the previous stages. For example, the mean duration of larvae of *E. hibisci* is:

$$\overline{D}_L = \frac{66 \times 8 + 78 \times (14-8) + 90 \times (18-14)}{8+6+4} - 57.33$$
$$= 75.33 - 57.33 = 18.00 \text{ h}$$

That is, from the time when the eggs were laid to the time when the larval stage ends totalled 75.33 h, in which the duration of the egg is 57.33 h and duration of the larva is 18 h.

Thus, we can calculate the mean duration of larva, nymph and total immature stages using the following formulas:

$$\overline{D}_L = \overline{D}_{(E \to N)} - \overline{D}_E \quad \cdots\cdots(5)$$

$$\overline{D}_N = \overline{D}_{(E \to A)} - \overline{D}_E - \overline{D}_L \quad \cdots\cdots(6)$$

$$\overline{D}_M = \overline{D}_{(E \to A)} = \overline{D}_E + \overline{D}_L + \overline{D}_N \quad \cdots\cdots\cdots\cdots\cdots\cdots\cdots\cdots\cdots\cdots\cdots\cdots (7)$$

where $\overline{D}_E$, $\overline{D}_L$, $\overline{D}_N$, $\overline{D}_M$ is the mean duration of eggs, larvae, nymphs and immature stages, respectively. $D_{(E \to N)}$ is the mean duration from oviposition to the nymph just appearing (i.e., larvae stage just ending). $D_{(E \to A)}$ is the mean duration from the oviposition to adult just appearing (i.e., nymph stage just ending), or the mean duration of the immature stage (Table 3).

Table 3. Comparison of two calculation methods on development stages of two species of phytoseiid mites

Groups	Eggs		Larvae		Nymph		Combined	
	Mean	SD	Mean	SD	Mean	SD	Mean	SD
E. hibisci								
isolated rearing	57.33	11.50	18.00	6.17	53.33	6.14	128.67	9.70
group rearing	57.33	11.50	18.00	9.70	53.33	9.70	128.67	9.70
E. stipulatus								
isolated rearing	53.29	2.91	21.18	5.25	51.69	7.57	126.00	10.95
group rearing	53.29	2.91	20.96	7.23	51.75	10.95	126.00	10.95

Discussion

If individuals, including eggs, larvae, and nymphs did not die and/or escape in the experiment, the same mean duration of eggs, larvae, nymphs and total immature stage will result from both types of calculation methods described above. The standard deviation will also be the same for the egg and total immature stage, but not for larval and nymphal stages. If some individuals of various stages happened to die and/or escape in the experiment, the mean duration of eggs, larvae and nymphs calculated by the method under conditions of group rearing could be either the same, more, or less, compared to the method under conditions of individual rearing. It is difficult to distinguish which possibility occurred because we do not know which individual died and/or escaped and what stage it was in. Using

E. stipulatus as an example (see right side of Table 2), at the 5th observation (60 h), there were 17 individuals, of which 13 were larvae and 4 nymphs. At the 6th observation there were 16 individuals, all of which were in nymphal stage. According to these two observations, one individual died and/or escaped. Generally, we can't distinguish from the record of group rearings in what stage the individual died and/or escaped. In the case of very low mortality and/or escape, however, the difference of the mean duration calculated by two methods is very small (Table 3).

The use of the group rearing may simulate a natural population and also is more practical, especially for experiments involving multiple treatments. However, the experimental method to use will depend on the purpose, conditions of the experiment, as well as biology of the experimental species.

Original article was published in *International Journal of Acarology*, volume 14, 1988

附录一
专家荣誉称号、获奖证书和授权专利目录

1. 专家荣誉称号

中华人民共和国农牧渔业部部属重点高等农业院校优秀教师(1985)
国务院政府特殊津贴专家(1992)
农业部有突出贡献的中青年专家(1992)
西南农业大学1994—1995年度优秀教师(1995)
重庆市教育委员会为人师表先进个人(1996)
全国高校科协工作研究会1996—1998年优秀工作者(1998)
重庆市优秀博士生指导教师(2005)
西南大学突出贡献奖(2009)

国务院学位委员会批准为博士生指导教师(1993)
四川省学位委员会第一届植保学科评议组成员(1995)
国务院学位委员会第四届土化植保学科评议组成员(1997)
重庆市学位委员会第一届生物学科评议组组长(2000)
重庆市植保学科学术技术带头人(2002)

农业部全国高等农业院校教学指导委员会农业昆虫小组组长(1997)
全国第二届农业病虫抗药性专家小组成员(1997—2003)
四川省第二届植物检疫对象审定委员会副主任委员(1998—2003)
农业部农技推广中心第二届全国农业病虫害专家组成员(1999—2003)
农业部第三、四届全国植物检疫性有害生物审定委员会委员(2003—2009)
西南大学发展规划专家咨询委员会委员(2009)
西南大学第一届研究生教育指导专家委员会委员(2009)

四川省科学技术顾问团第一、二届成员(1990、1994)
重庆市科学技术顾问团第一、二届成员(1998、2002)
涪陵地区科学技术顾问团第一届成员(2000)

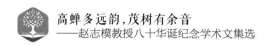

《西南农业大学学报》第五届编辑委员会成员(1994—1998)
《植物医生》第一届主编(1998)
《昆虫知识》第八届编辑委员会委员(2001)、第九届编辑委员会荣誉编委(2004)
Systematic & Applied Acarology 第一届编辑委员(2004)

重庆市农学会常务理事(1994)
重庆市植保学会秘书长、理事长、荣誉理事(1989—2001)
中国昆虫学会蜱螨专业委员会委员(1998—2002)
中国植物保护学会第九届理事、科学普及工作委员会副主任(2005)
重庆市第一届农药协会副理事长(2001)
重庆市生态学会第一、二、三届副理事长,名誉理事(2008)

中华人民共和国国家民族事务委员会在少数民族地区长期从事科技工作荣誉证书(1983)
中国人民政治协商会议重庆市第十届委员会委员(1993)

2. 科研教学奖励

中华人民共和国农牧渔业部农牧渔业技术改进一等奖(1984)——四川省农业害虫和天敌资源调查研究
中华人民共和国农牧渔业部科学技术进步三等奖(1986)——柑桔叶螨种群生态系统研究
农业部科学技术进步三等奖(1990)——桔园生态系昆虫群落及其控制
农业部科学技术进步三等奖(1994)——菜地昆虫群落及害虫综合防治研究
农业部科学技术进步三等奖(1997)——柑桔病虫综合防治新技术研究

四川省科学技术进步二等奖(1984)——四川省农业害虫和天敌资源调查研究
四川省科学技术进步二等奖(1991)——柑桔园昆虫群落及害虫综合治理研究
四川省计算机优秀软件一等奖(1991)——柑桔害螨综合管理决策支持系统研究
四川省科学技术进步一等奖(1993)——水稻丰收菌研究与应用
四川省科学技术进步二等奖(1993)——柑桔害螨综合管理决策支持系统研究
四川省科学技术进步三等奖(1996)——四川省主要农作物病虫害抗药性监测与研究

重庆市科学技术进步二等奖(1999)——新传入检疫对象——美洲斑潜蝇的研究与防除
重庆市科学技术进步三等奖(1999)——角倍林主要病虫种类及角倍丰产技术研究
重庆市北碚区科学技术进步奖一等奖(2005)——玉米地昆虫群落及优势种的成灾规律研究
重庆市自然科学三等奖(2009)——朱砂叶螨适应酸雨和杀螨剂胁迫的机理及适合度评估

西南农业大学重大科学成果一等奖(1985)——《生态学引论——害虫综合防治的理论及应用》
西南农业大学教学成果评比三等奖(1992)——柑桔害虫多媒体课件
西南大学科技二等奖(1993)——温光湿培养箱的设计与制作
西南农业大学优秀电教教材评比二等奖(1997)——柑橘害虫及防治管理
西南农业大学优秀教材评比二等奖(2003)——《农产品储运保护学》

3. 授权专利

何林,臧延琴,王进军,赵志模,于彩虹. 10%甲氰·阿维微乳剂配方. ZL 200510012041.3
赵志模,何林,宫庆涛,唐松,武可明. 柑橘大实蝇食物诱杀剂. ZL 201210050296.9
赵志模,何林,宫庆涛,唐松,武可明. 一种便于害虫监测和防治的诱捕器. ZL 201220073769.2
刘浩强,李鸿筠,冉春,姚廷山,胡军华,王进军,赵志模. 实蝇类昆虫饲养装置. ZL 201220002960.8
刘浩强,李鸿筠,冉春,姚廷山,胡军华,王进军,赵志模. 植物源杀虫剂. ZL 201210076671.7

附录二
专著、教材、论文目录

1. 专著与教材

赵志模(编著).《生态学引论——害虫综合防治的理论及应用》.科学技术文献出版社重庆分社,1984.

赵志模(副主编).《四川农业害虫天敌图册》.四川科学技术出版社,1990.

赵志模(副主编).《四川农业害虫及其天敌名录》.四川科学技术出版社,1986.

赵志模(编著).《群落生态学原理与方法》.科学技术文献出版社重庆分社,1990.

赵志模(主编).《农产品储运保护学》.中国农业出版社,2001.

赵志模(主编).《统筹城乡发展与植保科技进步》.中国农业科学技术出版社,2008.

赵志模(副主编).《昆虫生态学与害虫预测预报》.科学出版社,2012.

赵志模(编委).《中国农作物病虫害(第三版)》中册第15单元储粮病虫害.中国农业出版社,2015.

赵志模(主编).《植物保护学院史》.西南师范大学出版社,2016.

2. 科研论文

中文发表(*为本选集刊印论文)

[1] 赵志模.稻田寄生蜂群落种-多度关系、多样性及群落排序的探讨.西南农学院学报,1982(3):13-23.

[2] 朱文炳,赵志模,张永毅.昆虫与环境.西农科技,1982(1):22-29.

[3]* 赵志模.昆虫寄生作用数学模式的探讨——以纵卷叶螟绒茧蜂(*Apanteles cypris* Nixon)和稻纵卷叶螟(*Cnaphalocrocis medinalis* Guenee)自然种群为例.西南农学院学报,1983(1):68-83.

[4] 朱文炳,赵志模,张永毅.四川省害虫天敌资源初步调查 I.姬蜂科、茧蜂科和蚜茧蜂科.西南农学院学报,1982(1):47-59.

[5] 朱文炳,张永毅,赵志模,叶辉.四川省害虫天敌资源初步调查 II.瓢虫科 Coccinellidae.西南农学院学报,1982(3):1-13.

[6] 朱文炳,赵志模,张永毅.四川省害虫天敌资源初步调查 III.小蜂总科细蜂总科肿腿蜂总科.西南农学院学报,1982(4):28-41.

[7] 朱文炳,赵志模,张永毅.四川省害虫天敌资源初步调查 Ⅳ.双翅目 Diptera.西南农学院学报,1983(3):51-57.

[8]* 赵志模,朱文炳,叶辉,李强.桔全爪螨(*Panonychus citri* McGregor)实验种群生命表的组建与分析.西南农学院学报,1985(3):30-39.

[9]* 赵志模,朱文炳,郭依泉.桔园昆虫群落演替初步研究.西南农学院学报,1985(3):135-143.

[10] 朱文炳,赵志模,张永毅.四川麦田蚜茧蜂初步研究.中国昆虫学会40周年会刊,1985:249

[11] 赵志模.重庆市北碚区稻田寄生蜂类群初步考察.昆虫天敌,1986,8(3):125-136.

[12] 游兰韶,熊漱琳,朱文炳,赵志模.中国绒茧蜂属小志(六).湖南农学院学报,1984(3):53-60.

[13] 朱文炳,叶辉,赵志模.桔全爪螨(*Panonychus citri* McG.)种群动态的系统分析及其模拟——Ⅰ.与系统模拟有关的生物学特性.西南农学院学报,1985(3):2-11.

[14] 叶辉,赵志模,朱文炳.桔全爪螨(*Panonychus citri* McG.)种群动态的系统分析及其模拟——Ⅱ.系统分析、模拟及施药模拟实验.西南农学院学报,1985(3):12-21.

[15] 叶辉,赵志模,朱文炳.德氏钝绥螨对桔全爪螨捕食作用研究简报.西南农学院学报,1985(3):147-149.

[16] 朱文炳,赵志模.柑桔介壳虫寄生蜂考查简报.西南农学院学报,1985(3):149-150.

[17] 郭依泉,赵志模,朱文炳.桔园昆虫群落季节格局研究.西南农业大学学报,1987,9(1):27-32.

[18] 吴仕源,赵志模,张永毅.自装温、光、湿控制培养室.西农科技,1987(4):26.

[19] F.Du Toit,赵志模.冬小麦上俄国麦蚜经济阈值的研究.国外农学.植物保护,1988(1):24-27.

[20] 李隆术,朱文炳,赵志模,郭依泉.桔园昆虫群落研究现状及进展.西南农业大学学报,1988,10(2):132-136.

[21] 朱文炳,郭依泉,赵志模.四川桔园昆虫群落营养结构研究——Ⅰ蚧类和螨类亚群落.西南农业大学学报,1988,10(2):137-143.

[22] 郭依泉,朱文炳,赵志模.桔园昆虫群落空间结构研究.西南农业大学学报,1988,10(2):144-149.

[23] 李隆术,朱文炳,赵志模,周新远,陈杰林.柑桔叶螨种群生态系统研究(综合摘要).西南农业大学学报,1988,10(2):159-162.

[24] 赵志模,朱文炳,叶辉,郭依泉.德氏纯绥螨对桔全爪螨捕食作用的研究.西南农业大学学报,1988,10(2):186-192.

[25] 赵志模.美国柑桔害虫治理概况.国外农学.植物保护,1988(2):1-6.

[26] 嵇庆才,赵志模,吴仕源.桔园食螨瓢虫的空间格局.西南农业大学学报,1991,13(1):116-120.

[27] 吴仕源,赵志模,张永毅,嵇庆才.害虫管理中的气温预测.西南农业大学学报,1991,13(2):5-10.

[28] 赵志模,刘映红,雷蕾,向光瞻,林清,张宗美,张昌伦.朱砂叶螨、侧杂食线螨在茄子、辣椒、豇豆上的空间分布型.西南农业学报,1991,4(3):86-90.

[29] 赵志模,郭依泉,朱文炳.柑桔叶吸汁性害虫生态位研究及竞争群划分//中国生态学会青年研究会,中国昆虫学会昆虫生态专业委员会.青年生态学者论丛(二).昆虫生态学研究,1992,2:238-243.

[30] 邓永学,李隆术,吴仕源,赵志模,郭依泉,曾正.CO_2气调对两种储粮害虫致死率的研究.西南农业大学学报,1992,14(1):5-8.

[31] 吴仕源,曾正,邓永学,李隆术,赵志模,郭依泉.气体浓度控制仪的制作及使用.西南农业大学学报,1992,14(2):57-61.

[32] 郭依泉,赵志模.群落食物网间的相似性测度.生态学杂志,1992,11(3):65-68.

[33] 赵志模,陈艳,吴仕源.不同食物对普通钝绥螨发育和繁殖的影响.蛛形报,1992,1(2):49-56.

[34] 刘映红,赵志模.柑桔叶片受害对桔全爪螨种群的反馈作用研究.昆虫生态学研究,1992,2:267-272.

[35] 赵志模,陈宇,李华荣,刘灼均,颜思齐.水稻丰收菌对常用杀虫、杀菌、除草剂的抗性研究.《植物有益微生物的研究和应用——水稻丰收菌》,科学技术文献出版社,1992.

[36] 赵志模,李华荣,陈宇,颜思齐.水稻、小麦丰收菌对玉米的作用初探.《水稻丰收菌》,科学技术文献出版社,1992.

[37] 杨大旗,胡声荣,肖建国,赵志模,颜思齐.水稻丰收菌大田试验、示范结果汇总.《水稻丰收菌》,科学技术文献出版社,1992.

[38] 陈艳,赵志模.普通钝绥螨的生物学特性.福建农学院学报,1993,22(2):188-192.

[39] 赵志模,陈艳,吴仕源.普通钝绥螨(*Amblyseius vulgaris*)对朱砂叶螨(*Tetranychus cinnabarinus*)捕食作用的研究.蛛形学报,1993,2(1):31-35.

[40] 邓永学,吴仕源,李隆术,赵志模,杨荣.植物芳香油配合气调对杂拟谷盗成虫致死率研究.西南农业大学学报,1993,15(3):55-58.

[41] 岿庆才,赵志模,李隆术.桔园内多种天敌对桔全爪螨的控制作用.生态学报,1993,13(2):107-114.

[42]* 赵志模,张肖薇.腐食酪螨对低氧高二氧化碳气调的抗性.蛛形学报,1993,2(2):126-128.

[43] 雷蕾,林清,刘映红,赵志模,张宗美,向光瞻.菜青虫对甘蓝为害损失及经济阈值模拟研究.西南农业学报,1993,6(3):75-79.

[44] 陈文龙,李隆术,赵志模,朱文炳.束管食螨瓢虫对不均匀分布猎物的捕食行为.上海农学院学报,1993,11(3):209-213.

[45]* 赵志模,刘映红,张昌伦.重庆市郊不同种植制度菜地昆虫群落结构研究.植物保护学报,1994,21(1):39-45.

[46] 陈文龙,赵志模,李隆术,朱文炳.束管食螨瓢虫对桔全爪螨的捕食作用研究.西南农业大学学报,1994,16(1):27-31.

[47] 陈文龙,李隆术,赵志模,岿庆才.桔园三种天敌对桔全爪螨的捕食作用.上海农学院学报,1994,12(1):9-15.

[48] 陈文龙,赵志模.束管食螨瓢虫抽样技术研究Ⅰ.空间分布及抽样技术.上海农学院学报,1994,12(1):35-41.

[49] 王进军,赵志模,吴仕源,邓永学,郭依泉.不同温度下气调对嗜虫书虱急性致死作用的研究.西南农业学报,1994,7(1):70-74.

[50] 张永毅,赵志模,吴仕源,邓兴平,郭依泉.柑桔害螨抗性综合治理.植保技术与推广,1994(2):23-24.

[51] 周晨曦,赵志模,郭依泉.仓库害虫侵染田间粮食研究.西南农业大学学报,1994,16(4):311-313.

[52] 林荣寿,邓新平,赵志模,彭丽年,张永毅,刘映红.四川省柑桔害螨抗药性监测与综合治理.西南农业学报,1994,7(3):75-81.

[53] 雷蕾,刘映红,赵志模.秋甘蓝不同生育期叶片受害损失模型研究.重庆农业科技,1994(1/2):20-23.

[54] 张永毅,赵志模,邓新平.应用高压注射器防治柑桔害虫药效试验.中国柑桔,1994(2):22-25.

[55] 赵志模.弘扬爱国主义精神,充分发挥无党派知识分子在改革开放中的作

用.重庆知识分子联谊会通讯,1994:7-8.

[56] 吴仕源,赵志模,刘牛,李媛.四川小麦穗期蚜虫发生程度预测研究.西南农业大学学报,1995,17(1):24-27.

[57] 邱明生,赵志模.角倍蚜秋季迁飞和生殖能力的研究.西南农业大学学报,1995,17(1):39-41.

[58]* 邓永学,吴仕源,赵志模.双低储粮虫螨群落组成研究.西南农业大学学报,1995,17(1):42-44.

[59] 周晨曦,赵志模.重庆市近郊农户储粮现状及害虫发生情况的初步调查.粮油仓储科技通讯,1995(2):6-9.

[60] 周晨曦,陈宏,赵志模,王进军.重庆市农户贮粮害虫研究.西南农业大学学报,1995,17(3):237-238.

[61] 邓永学,朱文炳,赵志模.食物对巴西豆象生长发育的影响.西南农业大学学报,1995,17(3):244-248.

[62]* 赵志模,吕慧平,张权炳.柑桔矢尖蚧一代幼蚧发生期数理统计预测.植物保护学报,1995,22(3):217-222.

[63] 刘映红,李隆术,赵志模.桔全爪螨与柑桔相互作用的研究.蛛形学报,1995,4(2):103-110.

[64] 邓永学,李隆术,赵志模,兰景华.菜豆和饭豆对巴西豆象发育影响比较研究.粮食储藏,1996,25(1):3-6.

[65] 郭凤英,邓新平,赵志模,李俐俐.朱砂叶螨羧酸脂酶最优测试条件的选择.蛛形学报,1996,5(1):54-61.

[66] 王进军,赵志模,郭依泉.嗜虫书虱的有效积温及气调致死作用研究.西南农业大学学报,1996,18(2):155-158.

[67] 王进军,赵志模,郭依泉,周亦红,周晨曦.温度、湿度对嗜虫书虱生长发育和繁殖的影响.植物保护学报,1996,23(2):147-151.

[68] 张永毅,邓新平,赵志模,刘映红.桔全爪螨对水胺硫磷和氧化乐果的抗性测定(简报).西南农业大学学报,1996,18(3):201.

[69] 王进军,周晨曦,周亦红,赵志模,郭依泉.不同温度下嗜虫书虱实验种群生命表.郑州粮食学院学报,1996,17(2):79-83.

[70] 陈文龙,顾振芳,李隆术,赵志模.束管食螨瓢虫研究及利用概述.昆虫知识,1996,33(5):304-306.

[71] 雷蕾,刘映红,赵志模,林清,张宗美.秋甘蓝不同生育期叶片模拟受虫害损失模型.西南农业学报,1996,9(3):42-47.

[72] 陈宏,刘映红,邓新平,赵志模.苦瓜叶片提取液对家蚕和菜青虫取食、生长和存活的影响.西南农业学报,1996,9(3):68-71.

[73] 张永毅,赵志模,刘怀.阿巴丁对桔全爪螨的田间药效.植物医生,1996,9(5):19.

[74] 涂建华,蒋凡,廖华明,赵志模,王进军.四川省农户贮粮设施及损害研究.西南农业大学学报,1996,18(6):609-612.

[75] 邱明生,赵志模,徐学勤,吴猛,耐阳,彭兴福,郑秀兰.角倍蚜与其寄主间营养关系的研究.西南农业大学学报,1996,18(6):613-617.

[76] 邓新平,郭凤英,赵志模.朱砂叶螨 Tetranychus cinnalarinus(Boisduval)蛋白质含量测定中的两种方法比较.河南职技师院学报,1997,25(2):31-35.

[77] 郭凤英,邓新平,赵志模,吕慧平.朱砂叶螨酸性磷酸酯酶测定条件的研究.植物保护,1997(4):5-7.

[78] 邱明生,赵志模,徐学勤,苟阳.角倍倍园主要生态因子对五倍子产量的影响.生态农业研究,1997,5(3):19-23.

[79]* 吴仕源,王进军,赵志模.CO_2和溴氰菊酯在不同温度下对嗜卷书虱毒性的相互影响.中国粮油学报,1997,12(6):5-9.

[80] 王进军,赵志模,李隆术.嗜卷书虱高二氧化碳抗性品系的选育研究.中国昆虫学会第六次全国代表大会暨学术讨论会论文摘要集,1997.

[81] 赵志模,罗华元.酸雨对作物-朱砂叶螨系统的影响及机理研究.中国昆虫学会第六次全国代表大会暨学术讨论会论文摘要集,1997.

[82] 张永毅,邓新平,赵志模.四种杀虫剂对柑桔矢尖蚧的田间药效试验.植物医生,1998,11(1):37.

[83] 彭炜,范京安,赵学谦,吴志平,李刚,赵志模,陈文龙,石万成,李建荣.四川省美洲斑潜蝇发生与综合防治初步研究.植物检疫,1998,12(3):135-138.

[84] 陈宏,李俊霞,赵志模.瓜类主要害虫及其天敌的典范相关分析.天津师大学报(自然科学版),1998,18(2):43-48.

[85] 丁伟,赵志模,邓新平.巴豆对昆虫的作用及应用前景.世界农业,1998(9):31-32.

[86] 陈宏,徐秋曼,黄树君,赵志模.瓜类主要害虫生态位研究及竞争群划分.华北农学报,1998,13(3):125-131.

[87] 赵光潜,赵志模,邱学东,左发双.美洲斑潜蝇幼虫为害与秋豇豆产量损失关系研究.植保技术与推广,1998,18(5):34-35.

[88] 王进军,赵志模,李隆术,邓新平,吴仕源.气调、红桔油及温度对嗜卷书虱熏

蒸作用的交互效应研究.中国粮油学报,1998,13(6):55-58.

[89] 陈文龙,赵志模,朱文炳,彭炜,吴志平,赵光潜,谢忠友,肖连康,黄建伟.四川地区美洲斑潜蝇的寄主种类.西南农业大学学报,1999,21(1):55-58.

[90] 彭炜,赵学谦,杨光超,李刚,蒋辉,范京安,赵志模,陈文龙,刘勇,唐仲明,赵虹,李建明.四川攀西地区斑潜蝇发生和综合防治研究.西南农业大学学报,1999,21(1):59-63.

[91] 邱明生,赵志模.角倍蚜冬寄主侧枝匐灯藓的生长特性研究.生态学杂志,1999,18(2):10-12.

[92] 王进军,赵志模.不同食物对嗜卷书虱发育和繁殖的影响.昆虫知识,1999,36(2):95-97.

[93] 丁伟,赵志模,王进军.气调储藏及贮粮害虫的抗气性.世界农业,1999(5):33-35.

[94] 邱明生,赵志模,李隆术.环境因子对角倍蚜秋迁蚜生殖和雌性蚜发育的影响.昆虫学报,1999,42(2):145-149.

[95] 郭凤英,赵志模,邓新平.朱砂叶螨抗药性与羧酸酯酶活性的关系.农药学学报,1999,1(2):91-93.

[96] 邱明生,赵志模.角倍蚜干母种群的空间格局及其形成机理研究.华东昆虫学报,1999,8(1):66-69.

[97] 王进军,赵志模,李隆术.嗜卷书虱的实验生态研究.昆虫学报,1999,42(3):277-283.

[98]* 郭凤英,赵志模.朱砂叶螨对不同农药抗药性发展趋势的研究.蛛形学报,1999,8(2):118-121.

[99] 张永毅,吴仕源,赵志模,王进军,邓新平.橘全爪螨春季高峰发生程度与气象因素的关系.昆虫知识,1999,36(5):283-285.

[100] 王进军,赵志模,李隆术.不同温度下气调及红桔油对嗜卷书虱的熏蒸作用研究.粮食储藏,1999,28(5):3-9.

[101] 赵志模.气调储粮及储粮害虫的抗气性//中国科学技术协会、浙江省人民政府.面向21世纪的科技进步与经济社会发展(上册).中国科学技术协会、浙江省人民政府,1999:1.

[102] 叶鹏盛,曾华兰,李琼芳,江怀仲,赵志模.棉铃虫对不同棉花品种的选择性研究(简报).西南农业大学学报,1999(5):97.

[103] 郭凤英,邓新平,赵志模.影响叶螨磷酸酯酶活性的四因子数学模型.昆虫学报,1999,42(4):364-371.

[104] 邱明生,王进军,赵志模,李隆术.辐射对亚洲玉米螟体内几种水解酶活力的影响.植物保护学报,1999,26(4):319-323.

[105] 丁伟,赵志模,周亦红.独特的昆虫化石——琥珀昆虫.昆虫知识,1999,36(6):350-352.

[106] 刘怀,赵志模,吕良琼,王帅.竹裂爪螨实验种群生命表研究.西南农业大学学报,1999,21(6):556-560.

[107] 罗其荣,邱明生,张卫,赵志模,李隆术.$^{60}Co\gamma$辐射对亚洲玉米螟体内保护酶活力的影响.华东昆虫学报,1999,8(2):21-25.

[108] 李剑泉,赵志模,侯建筠.稻虫生态管理.西南农业大学学报,2000,22(6):496-500.

[109] 丁伟,周亦红,赵志模.斑潜蝇抗药性的几种监测方法.植物检疫,2000,14(2):80-83.

[110] 吕慧平,赵志模.矢尖蚧黄蚜小蜂和花角蚜小蜂生物学和生态学特性研究.昆虫学研究进展,1999,212-215.

[111] 郭凤英,赵志模,邓新平.朱砂叶螨抗药性与羧酸酯酶活性的关系.农药学报,1999,1(2):91-93.

[112] 李隆术,赵志模.我国仓储昆虫研究和防治的回顾与展望.昆虫知识,2000,37(2):84-88.

[113] 赵志模.我国柑桔害虫的研究现状.昆虫知识,2000,37(2):110-116.

[114] 丁伟,周亦红,赵志模.斑潜蝇抗药性的几种监测方法.植物检疫,2000,14(2):80-83.

[115] 丁伟,吴文君,赵志模.抗辛硫磷及氰戊菊酯的棉铃虫品系对苦皮藤素的敏感性.西南农业学报,2000,13(1):75-78.

[116] 王进军,赵志模,李隆术.嗜卷书虱的实验生态研究.广西植保,2000,13(2):4.

[117] 周亦红,赵志模,邓新平.美洲斑潜蝇和南美斑潜蝇幼虫分龄的研究.西南农业大学学报,2000,22(4):339-341.

[118] 刘怀,赵志模,丁伟,吕良琼,王帅,谭详国.长宁竹区竹裂爪螨生物学及楠竹害螨消长动态研究.四川林业科技,2000,21(3):1-4.

[119] 叶鹏盛,曾华兰,赵志模,李琼芳,江怀仲.形态抗虫棉对棉铃虫实验种群发育的影响研究.棉花学报,2000,12(5):250-253.

[120] 丁伟,赵志模,贾明芳,王建新,贾建平.区域性无公害农药应用技术体系的建立及展望.农业现代化研究,2000,21(5):313-316.

[121] 周亦红,赵志模,邓新平.温度对美洲斑潜蝇实验种群的影响.西南农业大学学报,2000,22(5):443-447.

[122] 周亦红,赵志模,邓新平,吴仕源.温度对南美斑潜蝇实验种群的影响.南京农业大学学报,2000,23(4):33-36.

[123] 周亦红,赵志模,邓新平.美洲斑潜蝇和南美斑潜蝇同龄幼虫种内竞争的研究.植物保护,2000,26(6):1-3.

[124] 李剑泉,赵志模,侯建筠.植保领域的蜘蛛研究进展.植物医生,2000,13(6):9-12.

[125] 叶鹏盛,曾华兰,李琼芳,江怀仲,赵志模.川棉243对棉铃虫的抗性及其机理.西南农业大学学报,2000,22(6):490-493.

[126] 李剑泉,赵志模,侯建筠.稻虫生态管理.西南农业大学学报,2000,22(6):496-500.

[127] 王进军,赵志模,李隆术.嗜卷书虱抗气调品系的选育及其适合度研究.昆虫学报,2001,44(1):67-71.

[128] 王进军,赵志模,李隆术.嗜卷书虱实验种群生命表的研究.应用生态学报,2001,12(1):83-85.

[129] 丁伟,赵志模,王进军,张永毅.高粱条螟在春玉米上的发生规律及危害特点.西南农业大学学报,2001,23(1):13-15.

[130] 李剑泉,赵志模,吴仕源.稻赤斑沫蝉的发生危害与防治对策.植物医生,2001,14(1):14-16.

[131] 刘怀,赵志模,王进军,徐学勤,李映平,张辅达.食物对竹裂爪螨生长发育及繁殖的影响.蛛形学报,2001,10(1):30-34.

[132] 李剑泉,赵志模,吴仕源,侯建筠.多物种共存系统中3种蜘蛛对褐飞虱的控制作用.蛛形学报,2001,10(1):35-40.

[133] 郭凤英,赵志模.抗药性叶螨的种群参数和相对适合度.蛛形学报,2001,10(1):41-43.

[134] 邱明生,张孝羲,王进军,赵志模.玉米田节肢动物群落特征的时序动态.西南农业学报,2001,14(1):70-73.

[135] 李剑泉,赵志模,吴仕源,明珂,侯丽娜.稻赤斑沫蝉的生物学与生态学特性.西南农业大学学报,2001,23(2):156-159.

[136] 陈宏,靳阳,徐秋曼,赵志模.瓜类蔬菜昆虫群落(包括蛛形纲和软体动物)的研究Ⅰ.群落的组成及结构.南开大学学报(自然科学版),2001,34(2):114-120.

[137] 刘怀,赵志模,吴仕源,徐学勤,李映平.竹裂爪螨实验种群密度效应研究.西南农业大学学报,2001,23(3):249-251.

[138] 张永毅,吴仕源,赵志模,邓新平,周亦红.柑橘矢尖蚧第一代发生期预测简报.植保技术与推广,2001,21(7):7-9.

[139] 周亦红,姜卫华,赵志模,邓新平.温度对美洲斑潜蝇及南美斑潜蝇种群增长的影响.生态学报,2001,21(8):1276-1284.

[140] 丁伟,李隆术,赵志模.书虱综合防治技术研究进展.粮食储藏,2001,30(4):3-6.

[141] 丁伟,王进军,赵志模.书虱实验种群饲养技术研究.西南农业大学学报,2001,23(4):304-306.

[142] 李剑泉,赵志模,朱文炳,侯丽娜,周彦,李雪燕.重庆市稻田动物群落及农田蜘蛛资源考察.西南农业大学学报,2001,23(4):312-316.

[143] 李剑泉,赵志模,侯建筠.稻田蜘蛛研究进展.蛛形学报,2001,10(2):58-63.

[144] 程伟霞,丁伟,赵志模.气调(CA)对储藏物害虫的作用机制.昆虫知识,2001,38(5):330-333.

[145]* 张智英,曹敏,杨效东,赵志模.舞草种子的蚂蚁传播.生态学报,2001,21(11):1847-1853.

[146] 丁伟,赵志模,王进军,李隆术.储藏物害虫书虱的综合治理//中国科协第四届青年学术年会重庆卫星会议暨重庆市第二届青年学术年会.重庆:西南师范大学出版社,2001,376-379.

[147] 邓永学,赵志模,李隆术.不同温度下食物对巴西豆象发育、繁殖及内禀增长率的影响//中国科协第四届青年学术年会重庆卫星会议暨重庆市第二届青年学术年会.重庆:西南师范大学出版社,2001,390-393.

[148] 邓永学,刘怀,赵志模,李隆术.储粮害虫防治措施探讨//城市昆虫学进展——全国第六届城市昆虫学术讨论会论文集.杭州:浙江大学出版社,2001:226-235.

[149] 丁伟,赵志模,李小珍.植物杀虫剂苦皮藤素对柑橘潜叶蛾控制效果的研究.植保技术与推广,2002,22(1):5-7.

[150] 丁伟,赵志模,王进军,朱文炳.玉米地节肢动物群落优势功能集团的组成与演替.生态学杂志,2002,21(1):38-41.

[151]* 李剑泉,赵志模.多物种共存系统中拟水狼蛛对三种稻虫的捕食作用.植物保护学报,2002,29(1):1-6.

[152] 李剑泉,赵志模,吴仕源,罗雁婕,明珂.多物种共存系统中蜘蛛对稻虫的控

制作用.中国农业科学,2002,35(2):146-151.

[153] 邓永学,赵志模,李隆术.高浓度CO_2气调防治谷蠹及杂拟谷盗的研究.粮食储藏,2002,31(1):3-6.

[154] 丁伟,王进军,赵志模,陈贵红.春玉米田蚜虫种群的数量消长及空间动态.西南农业大学学报,2002,24(1):13-16.

[155] 何林,赵志模,邓新平,王进军,刘怀,吴仕源.甲氰菊酯与阿维菌素单用、轮用和混用对朱砂叶螨抗性进化的影响.蛛形学报,2002,11(1):54-57.

[156] 丁伟,SHAAYA E,王进军,赵志模,高飞.两种昆虫生长调节剂对嗜虫书虱的致死作用.动物学研究,2002,23(2):173-176.

[157] 刘怀,赵志模,邓永学,何林,王进军,吴仕源.温度对竹盲走螨实验种群数量消长的影响.动物学研究,2002,23(4):356-360.

[158]* 邓新平,何林,赵志模.温度对朱砂叶螨二种抗药性品系发育和繁殖的影响.蛛形学报,2002,11(2):94-98.

[159] 李剑泉,沈佐锐,赵志模,罗雁婕.拟水狼蛛的生物学生态学特性.生态学报,2002,22(9):1478-1484.

[160] 陈宏,陈波,靳阳,赵志模.瓜类蔬菜昆虫群落(包括蛛形纲和软体动物)的研究Ⅱ.群落的数量动态.南开大学学报(自然科学版),2002,35(3):107-110.

[161] 何林,赵志模,邓新平,王进军,刘怀,刘映红.朱砂叶螨对三种杀螨剂的抗性选育与抗性风险评估.昆虫学报,2002,45(5):688-692.

[162] 丁伟,赵志模,王进军,程伟霞,李隆术.高CO_2对嗜卷书虱的致死作用及其行为反应.西南农业大学学报,2002,24(5):398-401.

[163] 张智英,李玉辉,赵志模.蚂蚁与蚁运植物的互惠共生关系.动物学研究,2002,23(5):437-443.

[164] 何林,赵志模,邓新平,王进军,刘怀.朱砂叶螨对两种杀螨剂的抗性遗传力及风险评估.植物保护学报,2002,29(4):331-336.

[165] 何林,邓新平,王进军,赵志模,陈冰勇,曹小芳.几种酰胺类除草剂毒力、药效及安全性评价.农药,2002,41(12):36-38.

[166] 程伟霞,王进军,王梓英,赵志模,丁伟.嗜卷书虱和嗜虫书虱酯酶性质的比较研究.农药学学报,2002,4(4):61-66.

[167] 任爽,邓新平,赵志模.栗瘿蜂生物学及综合防治技术研究进展.西南农业大学学报,2002,19(增刊):30-32

[168] 丁伟,赵志模,黎阳燕.丝棉木金星尺蠖的生物学特性及防治技术.植物保护,2002,28(3):29-31.

[169] 邓永学,赵志模,李隆术.环境因子对储粮害虫影响的研究进展.中国粮油学会第二届学术年会论文选集(综合卷),2002,37-43.

[170] 邓新平,何林,赵志模.30%虫螨净烟雾剂对桔蚜和桔全爪螨的控制作用.农药,2003,42(1):21-22.

[171] 丁伟,赵志模,肖崇刚.我国农药学科发展的现状、问题和对策.高等农业教育,2003(1):53-56.

[172] 张智英,李玉辉,赵志模.伊大头蚁蚁巢的结构与分布.昆虫学报,2003,46(1):40-44.

[173] 邓永学,赵志模,李隆术.环境因子对储粮害虫影响的研究进展.粮食储藏,2003,32(1):5-12.

[174] 何林,赵志模,邓新平,王进军,刘怀,刘映红.朱砂叶螨对3种杀螨剂的抗性选育及抗性治理研究.中国农业科学,2003,36(4):403-408.

[175] 丁伟,赵志模,王进军,董全雄,冯俊波.储粮环境中书虱猖獗发生的因子分析.粮食储藏,2003,32(2):12-17.

[176] 邓永学,赵志模,刘怀,王进军,何林.气调及温度对杂拟谷盗发育和繁殖的影响.植物保护学报,2003,30(2):217-218.

[177] 刘怀,赵志模,邓新平,邓永学,何林,吴仕源.竹裂爪螨自然种群消长及空间格局研究.蛛形学报,2003,12(1):32-37.

[178] 邓新平,何林,刘怀,赵志模.杀·苄粉剂对水稻二化螟和稻田杂草的控制作用.西南农业大学学报(自然科学版),2003,25(3):227-229.

[179] 陶卉英,丁伟,王进军,赵志模,张永强.嗜卷书虱抗气性与抗药性品系间羧酸酯酶活性的比较.西南农业大学学报(自然科学版),2003,25(4):349-352.

[180] 张智英,李玉辉,赵志模.云南热带亚热带退耕山地中蚂蚁对舞草种子的影响研究.林业科学,2003,39(5):74-77.

[181]* 丁伟,陶卉英,张永强,王进军,赵志模.磷化氢熏蒸处理对嗜卷书虱不同虫态的致死作用.农药学学报,2003,5(3):24-30.

[182] 丁伟,赵志模,王进军,陈贵红.三种玉米蚜虫种群的生态位分析.应用生态学报,2003,14(9):1481-1484.

[183] 丁伟,王进军,赵志模,李小珍.储藏物害虫嗜卷书虱对DDVP熏蒸的行为反应与致死剂量.应用生态学报,2003,14(9):1588-1590.

[184] 张永强,丁伟,赵志模,王进军,陶卉英.嗜卷书虱抗气性和抗药性品系呼吸代谢的比较.西南农业大学学报(自然科学版),2003,25(5):413-416.

[185] 丁伟,张永强,陈仕江,赵志模,朱聿振.14种中药植物杀虫活性的初步研究.

西南农业大学学报(自然科学版),2003,25(5):417-420.

[186] 张建萍,王进军,赵志模,何林,豆威.几种杀虫剂对朱砂叶螨酯酶的抑制作用.蛛形学报,2003,12(2):95-99.

[187] 程伟霞,王进军,赵志模,丁伟.嗜卷书虱和嗜虫书虱的研究进展.粮食储藏,2003,32(6):3-7.

[188] 吴仕源,赵志模,刘从军,陈新.《柑桔害虫》多媒体软件研制简报.植物医生,2003,16(6):23.

[189] 何林,谭仕禄,曹小芳,赵志模,邓新平,王进军.朱砂叶螨的抗药性选育及其解毒酶活性研究.农药学学报,2003,5(4):23-29.

[190] 丁伟,赵志模,王进军,陈贵红.三种玉米蚜虫种群的生态位分析.应用生态学报,2003,14(9):1481-1484.

[191] 赵志模.我国农作物病虫害综合防治的成绩和新形式下植保工作的任务.植物医生,2003(增刊):1-4.

[192] 赵志模.植保工作系统观与农作物病虫综合治理策略.植物医生,2003(增刊):5-8.

[193] 何林,曹守喜,臧延琴,赵志模.从农药剂型加工谈农药的无公害化.植物医生,2003(增刊):99-103.

[194] 张建萍,王进军,赵志模.频振式杀虫灯在无公害农产品生产中的作用.植物医生,2003(增刊):96-98.

[195] 张永强,丁伟,赵志模.植物性无公害杀螨剂研究.植物医生,2003(增刊):77-79.

[196]* 刘怀,赵志模,邓永学,徐学勤,李映平.温度对竹盲走螨实验种群生长发育与繁殖的影响.林业科学,2004,40(1):117-122.

[197]* 张建萍,王进军,赵志模.酸雨对植物—害虫—天敌系统的作用.昆虫知识,2004,40(1):11-15.

[198] 张永强,王进军,丁伟,赵志模.昆虫热休克蛋白的研究概况.昆虫知识,2004,40(1):16-19.

[199] 刘怀,赵志模,邓永学,邓新平,何林,吴仕源,帅霞.竹裂爪螨毛竹种群与慈竹种群对不同寄主植物的适应性及其生殖隔离.应用生态学报,2004,15(2):299-302.

[200]* 陈宏,靳阳,赵志模.瓜类蔬菜昆虫群落(包括蛛形纲和软体动物)的研究Ⅲ.群落的数量动态.南开大学学报(自然科学版),2004,37(1):92-96.

[201] 李小珍,刘映红,赵志模,周利飞.二点叶蝉自然种群的时空动态.动物学研

究,2004,25(3):221-226.

[202] 程伟霞,王进军,赵志模,丁伟,王梓英.四种杀虫剂对两种书虱羧酸酯酶和乙酰胆碱酯酶的抑制作用.动物学研究,2004,25(4):321-326.

[203] 张永强,丁伟,赵志模,王进军,刘丽红.中药植物丁香杀虫杀螨活性研究.西南农业大学学报(自然科学版),2004,26(4):429-432.

[204] 丁伟,赵志模,王进军,陶卉英,张永强.嗜卷书虱气调抗性与熏蒸剂抗性的相互关系.中国农业科学,2004,37(9):1308-1315.

[205] 张永强,丁伟,赵志模,王进军,廖涵杰.姜黄对朱砂叶螨的生物活性.植物保护学报,2004,31(4):390-394.

[206]* 何林,杨羽,符建章,王进军,赵志模.朱砂叶螨阿维菌素抗性品系选育及适合度研究.植物保护学报,2004,31(4):395-400.

[207] 曹小芳,何林,赵志模,邓新平,王进军.朱砂叶螨不同抗性品系酯酶同工酶研究.蛛形学报,2004,13(2):95-102.

[208] 张建萍,王进军,赵志模,陈洋,豆威.模拟酸雨对朱砂叶螨寄主植物三月早茄生理生化的影响.应用生态学报,2005,16(3):450-454.

[209] 尹勇,刘怀,赵志模,徐建华,罗林明.运用传播学原理比较传统农技推广与农民田间学校.西南农业大学学报(社会科学版),2005,3(1):1-4.

[210] 何林,赵志模,曹小芳,邓新平,王进军.朱砂叶螨抗甲氰菊酯品系选育及遗传分析.中国农业科学,2005,38(4):719-724.

[211] 何林,赵志模,曹小芳,邓新平,王进军.温度对抗性朱砂叶螨发育和繁殖的影响.昆虫学报,2005,48(2):203-207.

[212] 杨帮,丁伟,赵志模,梁志敏.美洲商陆和姜黄提取物抑菌活性的研究.西南农业大学学报(自然科学版),2005,27(3):297-300.

[213] 程绪生,刘怀,赵志模.抛秧稻田节肢动物及杂草群落结构研究.西南农业大学学报(自然科学版),2005,27(6):847-850.

[214] 张建萍,王进军,赵志模,陈洋,豆威.模拟酸雨对朱砂叶螨寄主植物三月早茄生理生化的影响.应用生态学报,2005,16(3):450-454.

[215] 白耀宇,赵志模,王进军,刘映红.转Bt基因作物及其对农田土壤节肢动物的影响.西南农业大学学报,2005,22(增刊):78-80.

[216] 刘怀,赵志模,王进军,吴仕源.竹盲走螨对竹裂爪螨的捕食功能.应用生态学报,2006,17(2):280-284.

[217] 丁伟,王进军,赵志模,张永强,陶卉英.嗜卷书虱抗气性和抗药性品系选育及抗性风险分析.植物保护学报,2006,33(1):81-86.

[218]　杨洪,王进军,赵志模,杨德敏,张宏.多次交配对松褐天牛精子数量消耗、产卵量和孵化率的影响.动物学研究,2006,27(3):286-290.

[219]　王文琪,王进军,赵志模.紫茎泽兰种子种群动态及萌发特性.应用生态学报,2006,17(6):982-986.

[220]*　王文琪,王进军,赵志模.不同微生境中泽兰实蝇寄生对紫茎泽兰有性繁殖的影响.植物保护学报,2006,33(4):391-395.

[221]　赵志模.发展中国家粮食产后损失(上).粮油仓储管理与科技信息,2006.

[222]　赵志模.发展中国家粮食产后损失(下).粮油仓储管理与科技信息,2006.

[223]　李晓芳,何林,赵志模.朱砂叶螨酯酶cDNA片段的基因克隆与序列分析.西南师范大学学报,2006,31(增刊):65-67.

[224]　刘开林,何林,缪应林,张云鹏,王进军,赵志模.高温和阿维菌素对朱砂叶螨的胁迫效应及热休克蛋白研究.中国农学通报,2007,23(1):249-253.

[225]　王文琪,王进军,赵志模.环境因素对紫茎泽兰种子萌发的影响.中国农学通报,2007,23(2):346-349.

[226]　刘开林,何林,王进军,赵志模.害虫及害螨对阿维菌素抗药性研究进展.昆虫知识,2007,44(2):194-200.

[227]　张永强,丁伟,吴静,赵志模.3种杀螨剂生物测定方法的比较研究.农药科学与管理,2007,28(7):50-53.

[228]　杨洪,王进军,赵志模,杨德敏,唐志强.松褐天牛的交配行为.昆虫学报,2007,50(8):807-812.

[229]　赵志模.重庆市植保事业的主要成就及新形势下面临的机遇与挑战.庆祝重庆市植物保护学会成立10周年暨植保科技论坛论文集,2007:1-4.

[230]　张磊,王文琪,王进军,赵志模.紫茎泽兰生殖生长期叶片矿质营养元素动态变化研究.中国农学通报,2007,123(12):306-310.

[231]　张永强,丁伟,赵志模,吴静,樊钰虎.不同生长时期黄花蒿提取物对朱砂叶螨的生物活性.生态学杂志,2007,26(12):1969-1973.

[232]　刘怀,赵志模,徐学勤,李映平.毛竹林竹冠节肢动物群落结构与多样性研究.林业科学研究,2007,20(6):841-846.

[233]　张永强,丁伟,赵志模.姜黄素类化合物对朱砂叶螨的生物活性.昆虫学报,2007,50(12):1304-1308.

[234]　张永强,丁伟,赵志模,吴静,樊钰虎.黄花蒿提取物对朱砂叶螨生物活性的研究.中国农业科学,2008,41(3):720-726.

[235]　段海,赵志模,何林,王晓军,姜辉,林荣华.20种杀虫剂对柑橘矢尖蚧的室内

[236] 沈丽,罗林明,陈万权,赵志模,王进军.四川省小麦条锈病流行区划及菌源传播路径分析.植物保护学报,2008,35(3):220-226.

[237] 赵志模.群落生态学研究在IPM中的作用及其面临的困境.植物保护科技创新与发展——中国植物保护学会2008年学术年会论文集,2008:3.

[238] 曹新民,赵志模,赵先平,王进军,邓永学,豆威.食物对四纹豆象生长发育的影响.西南大学学报(自然科学版),2008,30(10):120-123.

[239] 何林,薛传华,赵志模,王进军.朱砂叶螨抗性品系不同温度下的相对适合度.应用生态学报,2008,19(11):2449-2454.

[240] 闫香慧,刘怀,赵志模,何恒果,代建平,程登发.水稻褐飞虱灯下发生期及种群数量动态分析.植物保护学报,2008,35(6):501-506.

[241] 曹新民,邓永学,赵志模,王进军,豆威.温度对四纹豆象生长发育与繁殖的影响.昆虫知识,2009,46(2):233-237.

[242]* 张永强,丁伟,田丽,赵志模.黄花蒿提取物对柑橘全爪螨的生物活性.中国农业科学,2009,42(6):2217-2222.

[243] 王文琪,赵志模,王进军,陶热.生物入侵生态学研究进展.安徽农业科学,2009,37(25):12153-12155.

[244] 闫香慧,赵志模,刘怀,谢雪梅,肖晓华,程登发.应用马尔可夫链理论对褐飞虱和白背飞虱发生程度的预测.生态学报,2009,29(11):5799-5806.

[245] 王文琪,王进军,赵志模,张伟.不同生境紫茎泽兰对生态群落中节肢动物多样性的影响.西南大学学报(自然科学版),2009,31(12):14-20.

[246] 杨洪,王进军,赵志模,杨德敏,唐志强.松褐天牛生物学特性的研究.贵州农业科学,2010,38(1):77-80.

[247] 闫香慧,赵志模,刘怀,肖晓华,谢雪梅,程登发.白背飞虱若虫空间格局的地统计学分析.中国农业科学,2010,43(3):497-506.

[248] 杨洪,王进军,赵志模,杨德敏,唐志强.重庆松墨天牛发生规律研究.中国森林病虫,2010,29(2):15-17.

[249] 王文琪,王进军,赵志模,张伟.紫茎泽兰入侵对弃耕荒地植物群落多样性的影响.华中农业大学学报,2010,29(3):300-305.

[250] 王文琪,赵志模,王进军,陶热.不同生境群落特征及对紫茎泽兰幼苗生长动态的影响.九江学院学报(自然科学版),2010(2):11-15.

[251] 赵志模.转基因植物并不是解决病虫问题的一劳永逸和唯一的手段.中国科学技术协会学会学术部.新观点新学说学术沙龙文集44:转基因植物与食品

安全,2010:2.

[252] 赵志模.食品安全并不等于粮食安全.新观点新学说学术沙龙文集44:转基因植物与食品安全,2010:7.

[253] 周浩东,裴强,闫香慧,刘怀,王泽乐,刘祥贵,赵志模.褐飞虱和白背飞虱与主要天敌时间生态位研究.西南师范大学学报(自然科学版),2010,35(5):80-86.

[254] 何恒果,赵志模,王进军.柑橘全爪螨Na^+-K^+-ATPase的生化毒理学特性研究.果树学报,2011,28(1):102-107.

[255] 何恒果,赵志模,闫香慧,王进军.桔全爪螨对阿维菌素和甲氰菊酯的抗性现实遗传力及风险评估.应用生态学报,2011,22(8):2147-2152.

[256] 曾卓华,刘洪,肖国民,黄左权,田军,肖晓华,杨忠武,邹祥明,范开举,赵志模.柑橘大实蝇防控技术的新突破.植物检疫,2011,25(5):36-39.

[257] 武可明,赵志模,何林,宫庆涛,唐松.柑桔大实蝇的发生及综合防治.中国南方果树,2011,40(5):73-75.

[258] 闫香慧,刘怀,赵志模,肖晓华,程登发.重庆秀山县褐飞虱和白背飞虱种群发生规律研究.植物保护,2012,38(1):128-132.

[259]* 唐松,宫庆涛,豆威,王进军,赵志模.温度、土壤含水量和埋蛹深度对柑橘大实蝇羽化的影响.植物保护学报,2012,39(2):137-141.

[260]* 宫庆涛,武可明,唐松,何林,赵志模.柑橘大实蝇羽化出土及橘园成虫诱集动态研究.生物安全学报,2012,21(2):153-158.

[261] 刘浩强,向可海,李鸿筠,冉春,胡军华,姚廷山,王进军,赵志模.自制植物源农药对柑桔大实蝇的毒力、化蛹及羽化的影响.中国南方果树,2012,41(5):48-49.

[262] 肖伟,武可明,宫庆涛,赵志模,何林.柑橘大实蝇直肠粗提物的信息素功能.中国农业科学,2013,46(7):1501-1508.

[263] 刘浩强,李鸿筠,冉春,胡军华,姚廷山,王进军,赵志模.橘小实蝇生物学特点及控制方法.中国园艺文摘,2013,29(9):1-6.

[264] 刘浩强,向可海,李鸿筠,冉春,胡军华,姚廷山,王进军,赵志模,王联英.温度对柑橘大实蝇幼虫、化蛹和羽化的影响.中国园艺文摘,2013,29(10):13-14.

[265] 唐松,赵志模,王进军,杨忠武,田军,刘晓明,范开举,袁文彬,裴强.重庆地区柑桔大实蝇系统监测研究.环境昆虫学报,2014,36(1):44-50.

[266] 宫庆涛,赵志模,孙瑞红,何林,肖伟,李素红.多种物质对柑桔大实蝇的电生

理活性及田间引诱效果评价.环境昆虫学报,2014,36(1):95-101.

[267]　刘浩强,张云飞,李鸿筠,冉春,胡军华,姚廷山,王进军,赵志模.基于RAPD和SRAP分子标记的柑桔大实蝇种群多态性及其亲缘关系研究.西南大学学报(自然科学版),2014,36(11):49-56.

[268]*　何恒果,闫香慧,王进军,赵志模.甲氰菊酯和阿维菌素对柑橘全爪螨的亚致死效应.应用生态学报,2016,27(8):2629-2635.

英文发表(* Selected papers for this essays)

[1]*　ZHAO Z M, MCMURTRY J A. Calculation of developmental duration of mites reared in groups compared to those reared in isolation. International journal of acarology, 1988, 14(3): 137-141.

[2]*　ZHAO Z M, MCMURTRY J A. Development and reproduction of three *Euseius* (Acari: Phytoseiidae) species in the presence and absence of supplementary foods. Experimental & applied acarology, 1990, 8: 233-242.

[3]　ZHAO Z M. The predation of *Amblyseius vulgaris* to *Tetranychius cinnabarinus*// Congress of Entolmol(Abstracts). Secretarial of XIX internal, 1992.

[4]　ZHAO Z M, CHEN Y. Development and reproduction of *Amblyseius vulgaris* on different foods//Congress of Entomol(Abstracts). Secretaril of XIX interenal, 1992.

[5]　YE H, ZHAO Z M. Life table of *Tomicus Piniperda*(L.)(Col., Scolytidae) and its analysis. Journal of applied entomology, 1995, 119(1-5): 145-148.

[6]　GUO F Y, DENG X P, ZHAO Z M. Finding the optimal conditions for assaying the activity of glutathion-S-aromatic group transferases in *Tetranychus cinnabarinus* (Acari: Tetranychidae). Systematic and applied acarology, 1997, 2: 63-70.

[7]　GUO F Y, ZHANG Z Q, ZHAO Z M. Pesticide resistance of *Tetranychus cinnabarinus*(Acari: Tetranychidae) in China: a review. Systematic and applied acarology, 1998, 3(1): 3-7.

[8]　WANG J J, ZHAO Z M, LI L S. Studies on bionomics of *Liposcelis entomophila* (Psocoptera: Liposcelididae) infesting stored products. Entomologia sinica, 1998, 5(2): 149-158.

[9]　YE H, ZHAO Z M, ZHU W B. Life table of *Panonychus citri* McG. with particular reference to temperatures. Journal of Yunnan university, 1998, 20(5): 367-369.

[10]* ZHAO Z M, LUO H Y, WANG J J. Effects and mechanisms of simulated acid rain on plant-mite interactions in agricultural systems. I. The direct effects of simulated acid rain on carmine spider mite, *Tetranychus cinnabarinus*. Systematic and applied acarology, 1999, 4(1): 83-89.

[11] WANG J J, ZHAO Z M, LI L S. Selection of resistance strains in *Liposcelis bostrychophila* Badonnel to CO_2-enriched atmospheres. Entomologia sinica, 1999, 6(1): 45-52.

[12] WANG J J, ZHAO Z M, LI L S. Some biochemical aspects of resistance to controlled atmosphere in *Liposcelis bostrychophila* Badonnel (Poscoptera: Liposcelididae). Entomologia sinica, 1999, 6(2): 178-186.

[13] WANG J J, ZHAO Z M, LI L S. Induced tolerance of the psocid, *Liposcelis bostrychophila* Badonnel (Psocoptera: Liposcelididae), to controlled atmosphere. International journal of pest management, 1999, 45(1): 75-79.

[14] DENG Y X, LI L S, ZHAO Z M. Development and reproduction of *Callosobruchus chinensis* (Coleoptera: Bruchidae) on four legume plant seeds. Proceedings of the 7th international working conference on stored product protection, 1999, 1: 107-108.

[15] WANG J J, ZHAO Z M, LI L S. Biochemical mechanisms of *Liposcelis botrychophila* Badonnel (Poscoptera: Liposcelididae) resistant to controlled atmosphere. Proceedings of the 7th international working conference on stored product protection, 1999, 1: 160-164.

[16] WANG J J, ZHAO Z M, LI L S. Resistance of psocid, *Liposcelis bostrychophila* Badonnel (Psocoptera: Liposcelididae) and the stability to controlled atmosphere. Proceedings of the 7th international working conference on stored product protection, 1999, 1: 697-701.

[17] WANG J J, ZHAO Z M, LI L S. Ecological fitness of CA resistant and susceptible strains of *Liposcelis bostrychophila* B. (Psocoptera: Liposcelididae). Proceedings of the 7th international working conference on stored product protection, 1999, 1: 702-705.

[18] WANG J J, ZHAO Z M, LI L S. Study on tolerance to oxygen deficiency, genetic

stability and ecological fitness of psocid, *Liposcelis bostrychophila* Badonnel (Psoceoptera: Iposcelididae). Zoological research, 1999, 20(2): 104-110.

[19] WANG J J, TSAI J H, ZHAO Z M, LI L S. Development and reproduction of the psocid *Liposcelis bostrychophila* (Psocoptera: Liposcelididae) as a function of temperature. Annals of the entomological society of America, 2000, 93(2): 261-270.

[20] WANG J J, ZHAO Z M, TSAI J H. Resistance and some enzyme activities in *Liposcelis bostrychophila* Badonnel (Psocoptera: Liposcelididae) in relation to carbon dioxide enriched atmospheres. Journal of stored products research, 2000, 36(3): 297-308.

[21] LIU H, ZHAO Z M, WANG J J, LI L S, DING W. Temperature-dependent development and reproduction of the spider mite, *Schizotetranychus bambusae* Reck (Acari: Tetranychidae). Systematic & applied acarology, 2000, 5(1): 33-39.

[22] GUO F Y, ZHAO Z M. Feeding behaviour of omethoate-resistant spider mites (Acari: Tetranychidae): a study using electrical penetration graphs. Systematic & Applied Acarology, 2000, 5(1): 3-7.

[23] WANG J J, TSAI J H, ZHAO Z M, LI L S. Interactive effects of temperature and controlled atmosphere at biologically relevant levels on development and reproduction of the psocid, *Liposcelis bostrychophila* Badonnel (Psocoptera: Liposcelididae). International journal of pest management, 2001, 47(1): 55-62.

[24] WANG J J, TSAI J H, DING W, ZHAO Z M, LI L S. Toxic effects of six plant oils alone and in combination with controlled atmosphere on *Liposcelis bostrychophila* (Psocoptera: Liposcelididae). Journal of economic entomology, 2001, 94(5): 1296-1301.

[25] LI J Q, SHEN Z R, ZHAO Z M. The predatory function of three spiders to two insect pests in rice within a multi-species co-existence system. Agricultural sciences in China, 2002, 1(4): 391-396.

[26] DING W, WANG J J, ZHAO Z M, TSAI J H. Effects of controlled atmosphere and DDVP on population growth and resistance development by the psocid, *Liposcelis bostrychophila* Badonnel (Psocoptera: Liposcelididae). Journal of stored products research, 2002, 38(3): 229-237.

[27]　　WANG J J, ZHAO Z M, DENG X P, DING W, LI L S. Comparison of energy reserves and utilization in *Liposcelis bostrychophila* populations selected for tolerance to controlled atmosphere. Entomologia sinica, 2002, 9(4):41-46.

[28]　　DING W, SHAAYA E, WANG J J, ZHAO Z M, TAO H Y. Efficacy of pyriproxyfen and methoprene for control of *Liposcelis bostrychophila* Badonnel (Psocoptera: Liposcelididae). Chinese journal of pesticide science, 2003, 5(4):15-22.

[29]　　HE L, ZHAO Z M, DENG X P, WANG J J, LIU H, LIU Y H. Resistance selection of *Tetranychus cinnabarinus* to three acarcides and its management strategy. Agricultural sciences in china, 2003, 2(2):183-189.

[30]　　HE L, ZHAO Z M, DENG X P, WANG J J, LIU H. Estimation of realized heritability of resistance to fenpropathrin, abamectin, pyridaben and their mixtures in acaricide-selected strains of *Tetranychus cinnabarinus* (Boiduval). Entomologia Sinica, 2003, 10(1):35-41.

[31]　　HE L, ZHAO Z M, DENG X P, WANG J J, LIU H. Resistance risk assessment: realized heritability of resistance to methrin, abamectin, pyridaben and their mixtures in the spider mite, *Tetranychus cinnabarinus*. International journal of pest management, 2003, 49(4):271-274.

[32]*　WANG J J, ZHAO Z M. Accumulation and utilization of triacylglycerol and polysaccharides in *Liposcelis bostrychophila* (Psocoptera, Liposcelididae) selected for resistance to carbon dioxide. Journal of applied entomology, 2003, 127(2):107-111.

[33]　　WANG J J, CHENG W X, DING W, ZHAO Z M. The effect of the insecticide dichlorvos on esterase activity extracted from the psocids, *Liposcelis bostrychophila* and *L. entomophila*. Journal of insect science, 2004, 4(1):1-5.

[34]　　WANG J J, ZHAO Z M, ZHANG J P. The host plant-mediated impact of simulated acid rain on the development and reproduction of *Tetranychus cinnabarinus* (Acari: Tetranychidae). Journal of applied entomology, 2004, 128(6):397-402.

[35]　　ZHANG J P, WANG J J, ZHAO Z M, DOU W, CHEN Y. Effects of simulated acid rain on the physiology of carmine spider mite, *Tetranychus cinnabarinus* (Boiduvals) (Acari: Tetranychidae). Journal of applied entomology, 2004, 128(5):

342-347.

[36] DING W, ZHAO Z M, WU W J, TAO H Y, WANG J J. Action mechanism and biological activity of celangulin to *Tetranychus cinnabarinus* (Acari: Tetranychidae). Systematic & applied acarology, 2004, 9(1): 27-32.

[37] HE L, ZHAO Z M, CAO X F, DENG X P, WANG J J. Resistance selection of *Tetranychus cinnabarinus* to fenpropathrin and genetic analysis. Agricultural science in China, 2005, 4(11): 851-856.

[38]* DOU W, WANG J J, ZHAO Z M. Toxicological and biochemical characterizations of GSTs in *Liposcelis bostrychophila* Badonnel (Psocop., Liposcelididae). Journal of applied entomology, 2006, 130(4): 251-256.

[39] WANG J J, ZHANG J P, HE L, ZHAO Z M. Influence of long-term exposure to simulated acid rain on development, reproduction and acaricide susceptibility of the carmine spider mite, Tetranychus cinnabarinus. Journal of insect science, 2006, 6(1): 1-8.

[40]* DONG P, WANG J J, ZHAO Z M. Infection by Wolbachia bacteria and its influence on the reproduction of the stored-product psocid, Liposcelis tricolor. Journal of insect science, 2006, 6(1): 1-7.

[41] WANG J J, ZHANG J P, DOU W, ZHAO Z M. Influence of simulated acid rain on population dynamics of Carmine Spider Mite, *Tetranychus cinnabarinus* (Boisduval) (Acari: Tetranychidae) and its host plant. International journal of acarology, 2008, 34(4): 427-434.

[42] HE L, GAO X W, WANG J J, ZHAO Z M, LIU N N. Genetic analysis of abamectin resistance in *Tetranychus cinnabarinus*. Pesticide biochemistry and physiology, 2009, 95(3): 147-151.

[43] HE L, XUE C H, WANG J J, LI M, LU W C, ZHAO Z M. Resistance selection and biochemical mechanism of resistance to two Acaricides in *Tetranychus cinnabarinus* (Boiduval). Pesticide biochemistry and physiology, 2009, 93(1): 47-52.

[44] YAN X H, LIU H, WANG J J, ZHAO Z M, HUANG F N, CHENG D F. Population forecasting model of *Nilaparvata lugens* and *Sogatella furcifera* (Homoptera: Delphacidae) based on Markov chain theory. Environmental entomology, 2010, 39

(6):1737-1743.

[45] HE H G, JIANG H B, ZHAO Z M, WANG J J. Effects of a sublethal concentration of avermectin on the development and reproduction of citrus red mite, *Panonychus citri* (McGregor) (Acari: Tetranychidae). International journal of acarology, 2011, 37(1):1-9.

[46] CHEN E H, DOU W, HU F, TANG S, ZHAO Z M, WANG J J. Purification and biochemical characterization of glutathione S-transferases in *Bactrocera minax* (Diptera: Tephritidae). Florida entomologist, 2012, 95(3):593-601.

[47] WANG J, ZHOU H Y, ZHAO Z M, LIU Y H. Effects of juvenile hormone analogue and ecdysteroid on adult eclosion of the fruit fly *Bactrocera minax* (Diptera: Tephritidae). Journal of economic entomology, 2014, 107(4):1519-1525.

附录三
荣誉和获奖证书(部分)

中华人民共和国国家民族事务委员会在少数民族地区
长期从事科技工作荣誉证书(1983)

中华人民共和国农牧渔业部农牧渔业技术改进一等奖(1984)

中华人民共和国农牧渔业部部属重点高等农业院校优秀教师(1985)

四川省科学技术进步奖二等奖荣誉证书(1991)

附录三　荣誉和获奖证书(部分)

农业部有突出贡献的中青年专家(1992)

国务院政府特殊津贴证书(1992)

四川省科学技术顾问团工作期间突出贡献奖(1994)

第二届全国高等农业院校教学指导委员会农业昆虫小组组长聘书(1997)

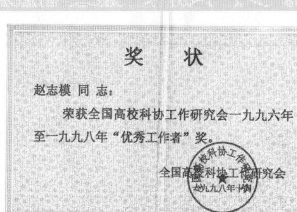

全国高校科协工作研究会1996—1998年优秀工作者奖状(1998)

重庆市科学技术进步二等奖(1999)

《昆虫知识》第八届编辑委员会委员聘书(2001)

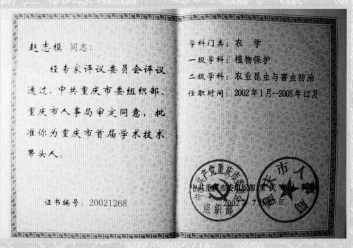

重庆市首届学术技术带头人(2002)

附录三 荣誉和获奖证书(部分)

国务院学位委员会第四届学科评议组重要贡献证书(2003)

全国第二届农业病虫抗药性专家小组重要贡献荣誉证书(2003)

重庆市学术技术带头人(2008)

西南大学突出贡献奖(2009)

附录四 工作生活掠影

全家福
赵先生六十岁生日合影

全家福

赵先生八十岁生日合影

赵先生八十寿诞庆典合影

附录四　工作生活掠影

人才培养 ▶

赵先生在操作286系统的电脑,在当时大大地提高了工作效率

1990年,在封昌远教授指导下调查柑橘病虫害情况

1990年,参加"水稻丰收菌使用效果试验与研究",与颜思齐教授一起考察水稻产量

赴四川省长宁县蜀南竹海考察

指导博士生郭凤英整理科研资料

准备教案、讲稿,上好每一堂课

2003年6月,赵先生与何林、方卫国、郭晓霞博士合影留念

2004年6月,在南充参加西华师范大学研究生答辩后,答辩委员会合影

赵先生仔细地查看答辩研究生的毕业论文

2005年6月,赵先生与西南农业大学(今西南大学)植物保护学院师生一起参加课外活动

2008年6月,赵先生参加西南大学2008届博士学位论文答辩后合影留念

2010年8月,与华中农业大学张国安先生和西北农林科技大学赵惠燕女士一起参加全国高等农业院校植保专业教材《昆虫生态与预测预报》编委会合影留念

2008年6月,赵先生参加西南大学植物保护学院2008届硕士生答辩的师生合影

学术会议 ▶

赵先生参加中美生物防治科学讨论会合影

1997年,部属高等农业院校教学指导委员会植保学科组专家合影

高蝉多远韵,茂树有余音
——赵志模教授八十华诞纪念学术文集选

1998年8月,全国第二届农业高校教学指导委员会植保学科会议合影(泰安)

1998年10月,全国第七届蜱螨学SAAS第一届国际学术讨论会合影

2002年,第八届《昆虫知识》全国编委会山东长岛会议代表合影

附录四　工作生活掠影

2003年,第三届全国植物检疫性有害生物审定委员会成立大会合影

2003年10-11月,赵先生参加国际仓储培训班的合影留念

2004年8月,参加国际外来生物入侵对策学术会,与会议主持人万方浩先生和部分代表合影

2004年10月，在中国昆虫生态学与害虫管理发展战略高级研讨会工作大会上发言

2004年10月，在中国昆虫生态学与害虫管理发展战略高级研讨会与参会代表合影

2007年12月，赵先生参加农作物病虫害预测及防治技术培训班合影（厦门）

附录四　工作生活掠影

参加全国有害生物防治学术会议(吉林),与会议主持人张帆女士及部分生防专家合影

2014年5月,中国生态学学会"农业有害生物监测预警"学术沙龙合影(湖北武汉)

考察交流

1987年，赵先生在美国加州大学做访问学者时与房东Allen&Mariley合影

赵先生受FAO邀请，访问菲律宾，考察农民田间学校

2002年参加《昆虫知识》编委会时，与主编杨星科先生合影

2003年3月，在广西参加全国农药会议后，赴越南考察时的留影

2003年10月，全国知名昆虫生态学家张孝曦先生访问西南农业大学（今西南大学）植保学院后与研究室部分研究人员合影

2003年10月，与全国著名隐翅虫分类专家郑发科先生亲切交谈

附录四　工作生活掠影

2003年11月,考察铜梁国库合影留念

2003年11月,与万方浩研究员进行交流

2003年11月,在四川省江油市考察大实蝇绿色防控的效果

2004年10月,访问扬州大学后,参观江南园林风景的合影

2004年12月,大雪纷飞的北京,赵先生受邀参加中国农科院植保所成果鉴定会,梁沛先生陪同参观了颐和园

2006年4月,山西植保站副站长钟灵女士陪同考察利用捕食螨控制害螨情况后,与基层植保技术人员合影

2006年4月，赵先生深入果园调查柑橘害虫的发生和为害

2006年7月，受重庆市林业局森保站邀请，与西南大学植物生态专家刘玉成先生一起在巫溪红池坝考察华山松死亡情况

2006年11月，农业部农技中心测报处处长张跃进先生访问西南大学植物保护学院后，与学院领导和研究人员合影

2007年4月，赵先生在柑橘园里调查吹绵蚧为害情况

2008年4月，在华南农业大学参加行业项目论证会后，参观深圳世纪公园

2008年4月，在华南农业大学参加行业项目论证会后，参观深圳世纪公园

附录四　工作生活掠影

2008年4月,原来的万县农校已合并为三峡职业学院,赵先生拜访自己的母校,并和副校长龙仕平先生合影

2008年4月,赵先生与课题组成员一起考察万县橘园

2008年4月,赵先生在田间仔细观察小麦蚜虫的为害情况

2008年6月,与知名两迁昆虫专家南京农业大学翟保平教授一起,考察指导重庆市秀山县的两迁害虫

2008年11月,赵先生与植保学会同仁参观香港理工大学

2009年4月,指导重庆市云阳县松树柏木叶蜂飞机防治工作,赵先生亲自乘直升机查看了山形和喷药情况

2009年6月，游武隆天坑地缝留影

2009年7月，在云阳考场诱集大实蝇的效果

2010年6月，仔细观察四川省江油市柑橘害虫发生情况

2011年9月，受邀访问甘肃农业大学的合影留念

2011年11月，与西南大学植保学院师生考察重庆市石柱县病虫防治工作合影留念

2011年11月，在苏州举行的全国植保大会上，与知名昆虫学家梁广文先生合影留念

附录四　工作生活掠影

2012年11月,考察重庆市巫山县病虫害发生情况,与西南大学植保院女教师代表们的合影

2014年4月,赵先生仔细查看大棚中黄板诱虫的效果

访问扬州大学后,游览江南风景园林

2014年11月,在福建泉州,与校友和全国青年创业者吴艺明合影

高蝉多远韵,茂树有余音
——赵志模教授八十华诞纪念学术文集选

附录五
指导的学生

博士后

孟庆繁

1998年博士后出站。北华大学教授,科研处处长。

博士

叶 辉

1992届,云南大学教授,原生命科学学院院长、农学院院长。

吕龙石

1993届,延边大学教授,原农学院副院长。

金道超

1995届,贵州大学教授,原副校长,贵州省人民政府参事。

罗华元

1995届,红云红河烟草(集团)公司研究员、首席农艺师。

王进军

1997届,西南大学教授,校长。

郭凤英

1998届,加拿大多伦多大学课程主管。

高蝉多远韵,茂树有余音
——赵志模教授八十华诞纪念学术文集选

邓永学

2001届,西南大学植物保护学院研究员。

李剑泉

2001届,中国林业科学研究院研究员。

刘　怀

2001届,西南大学教授,植物保护学院院长。

张智英

2001届,云南大学生态学与环境学院教授。

附录五 指导的学生

丁 伟

2002届,西南大学教授,农药研究所所长。

何 林

2003届,西南大学教授,发展规划与学科建设部部长、学术委员会办公室主任。

张建萍

2004届,石河子大学农学院教授。

杨 洪

2006届,贵州大学农学院教授。

王文琪

2006届,九江学院旅游与国土资源学院副教授。

张永强

2008届,西南大学副教授,原植物保护学院农药学系主任。

沈　丽

2008届,农业推广研究员,四川省种子站站长。

闫香慧

2010届,西华师范大学生命科学学院副教授。

何恒果

2010届,西华师范大学生命科学学院副教授。

赵云芬

2011届,西南大学教授,原法学院副院长。

硕 士

涂建华

1990届,农业推广研究员,四川省人民政府参事,四川省农业厅一级巡视员,副厅长。

范青海

1991届,新西兰第一产业部植物健康与环境实验室首席科学家。

陈 艳

1991届，新西兰初级产业部植物健康与环境实验室工作。

张肖薇

1993届，成都瑞耀国际贸易有限公司总经理。

吕慧平

1995届，浙江大学档案馆副研究馆员、干部人事档案室副主任。

陈 宏

1996届，天津师范大学教授，原生命科学学院副院长。

附录五　指导的学生

陈文龙

1997届,贵州大学教授,农学院昆虫研究所副所长。

叶鹏盛

1999届,研究员,四川省农业特色植物研究院党委书记。

周亦红

1999届,美国佐治亚大学工作。

程绪生

2000届,重庆市万州区农技站高级农艺师。

高蝉多远韵,茂树有余音
——赵志模教授八十华诞纪念学术文集选

曾小芳

2003届,西安交大阳光小学教师。

尹 勇

2004届,农业推广研究员,四川省农业厅植物保护站站长。

刘 洪

2005届,农业推广研究员,原重庆市植物保护站副站长。

薛传华

2008届,上海树农化工有限公司副总经理。

宫庆涛

2012届,山东省果树研究所助理研究员。

唐松

2012届,重庆市梁平区农业技术服务中心副主任,农艺师。

武司明

2013届,重庆市璧山区农业技术推广中心植保种子科科长。

后 记

经过近年的酝酿、筹划和收集资料,今年的紧密协商和紧张编排,这本涵盖我们敬爱的导师赵志模先生求学经历、职业生涯、学术成就和人才培养等内容的学术论文选集终于要出版了,我们感到既欣慰,又有些遗憾和愧疚。2019年4月18日是赵先生80诞辰,当时我们在校的几个弟子与校外的同门共同商定,一是要召开一次以先生学术思想为主线的学术沙龙,二是要出版一本先生的论文选集。学术沙龙如期举办,文选出版却由于疫情等原因一拖再拖。赵先生今年已是85岁,希望这份迟到的礼物能够表达我们的歉意,获得先生的原谅。书稿付梓,回首初心,我们执着于出一本赵先生的论文选集,除了要借此表达对恩师悉心栽培和谆谆教诲的感激之情外,更是要践行先生求真务实之精神,坚持一种传承,坚守一份责任:2000年11月29日,是先生的导师李隆术教授80诞辰,那一年,61岁的赵先生经常在办公桌前一坐一整天,默默无闻、一丝不苟地整理着李隆术教授的论文,是年10月,《储藏物昆虫和农业螨类研究——李隆术论文选》一书出版,这是先生给导师最好的生日礼物!先生已如此,学生当如何?

赵志模先生因其务实的作风、谦虚的为人和精深的学术造诣,在我国植保学界享有崇高的声誉。他为我国昆虫生态学、农产品储运保护学和螨类毒理学领域培养了大批优秀人才,包括数十名硕士和博士,他们在各自岗位上取得的点滴成绩,无不凝聚着先生的心血。为了继承和发扬先生的学术思想,总结他的奋斗历程和学术成就,表达对恩师的感激之情、祝贺之意,他的弟子们筹资出版了这本论文选集。

在这本论文选集谋划之初，我们就商定不单单是要出版一本反映先生学术成就的论文选集，更希望通过这本论文选集也能反映出先生从求学到退休大半生的学习、工作和生活经历，以期勾勒出鲜活的先生形象。现在看来，这个决定是正确的，我们不仅能从此书中继续汲取先生的学术营养，惊叹其学术成就、受到鼓舞，还能从中一窥他与时代发展同频共振的成长史、奋斗史，更能从中感悟出为学之道、为师之道、为人之道。赵志模先生不仅是我们的授业恩师，更是我们的人生导师。这本论文选集的主体部分收录先生已发表论文中的一小部分，论文选取征得他同意，未收录论文也以附录形式于书后全部呈现，便于查阅。论文收录分成昆虫生态学与害虫综合治理研究、农产品储运保护研究和螨类毒理学研究3个部分，各部分按论文发表时间和中英文的先后顺序排列。为了尊重历史，除对文中个别之处做适当修正外，收录论文基本上保持原貌。论文选集中的各个部分均经先生亲自审定，附录部分的荣誉获奖证书、工作生活掠影照片由赵先生及其家属提供，其他附录材料由编委会收集整理。

该书的出版，得到了众多科教工作者、院士和专家、赵志模先生家人及朋友等的大力支持，得到了西南大学、植物保护学院、文学院、发展规划与学科建设部、出版社领导和专家等的关怀。植物保护学院的豆威教授、王梓英副教授、徐志峰副教授和李金航博士后做了大量编撰和校对工作。挂一漏万，在此一并对大家的关心帮助表示衷心的感谢！

由于时间紧迫和其他因素限制，书中难免存在不足和疏漏之处，恳请读者谅解并批评指正。

编写组

二〇二四年 仲夏

赵志模教授致编写组的亲笔信

德昌县文化体育影视新闻出版局用笺

编辑组及全体成员：

收到你们精心编撰的《高弹多虑韵，故书抒余意——赵志模教授八十华诞纪念暨学术论文集》书稿，我倍感欣慰。在此，我谨向你们表示最衷心的感谢和崇高的敬意！

首先，我要感谢你们为编写这本文集所付出的心血。这份珍贵的礼物让我深感荣幸，也让我看到你们对我的这份情谊，我将永远铭记在心。

其次，我想说，我对你们在学术研究上的指导、帮助，以及在思想、生活上的关心，这是我作为一个教师的本分。你们在工作中取得的成绩，离不开自己的勤奋努力和组织的精心培养。我为你们的成长和进步感到由衷的高兴，也

第　　页

德昌县文化体育影视新闻出版局用笺

期待你们在未来的学术道路上、工作上取得更加辉煌的成就。

最后，我深感你们在文集中对我的赞誉太过。我所取得的学术研究和教书育人的成绩，是我们团队——昆虫生态与螨类综合治理部级重点实验室的全体同仁共同努力的结果，其中也包括研究生们。你们把许多荣誉归功于我，我倍感惭愧。在此，我要感谢团队中的每一位成员，是你们的智慧和汗水铸就了今天的成绩。我也想把这本文集献给我尊敬的恩师：李隆术、王辅和朱文炳教授。并衷心感谢曾经帮助过我的其他老师。

再次感谢编写组及研究生们为我送上的这份厚礼！愿我们继续保持美好的师

德昌县文化体育影视新闻出版局用笺

生情谊，共同为我国的植物医学事业贡献力量。祝愿编写组的全体成员及研究生们学业有成、事业顺利、生活幸福！

谨致 敬礼！

赵志模
于凉山德昌
2014.8.10